Crimes
Against
Nature

ALSO BY ROBERT F. KENNEDY, JR.

The Riverkeepers: Two Activists Fight to Reclaim Our Environment as a Basic Human Right

Crimes Against Nature

How George W. Bush and His Corporate Pals Are Plundering the Country and Hijacking Our Democracy

ROBERT F. KENNEDY, JR.

HarperCollins*Publishers*

HarperCollins books may be purchased for educational, business, or sales promotional use. For information, please write: Special Markets Department, HarperCollins Publishers Inc., 10 East 53rd Street, New York, NY 10022.

FIRST EDITION

Designed by Jaime Putorti

Printed on acid-free paper

Library of Congress Cataloging-in-Publication Data is available upon request.

ISBN 0-06-074687-4

04 05 06 07 08 ❖/RRD 10 9 8 7 6 5 4 3 2 1

To my wife, Mary Richardson,
who encourages me with her love, patience, and faith;
inspires me with her energy; and is the best
environmentalist in our family

Contents

Crimes Against Nature

Introduction

Earlier this year I was invited to speak at the Round Hill Club in Greenwich, Connecticut. If Greenwich is the Republicans' Mecca, then the Round Hill Club is the Kaaba. In the foyer I passed beneath an oversized photograph of Senator Prescott Bush, a former Greenwich resident and the current president's grandfather. Somebody pointed to an anteroom and commented: "That's where George met Barbara," referring to the president's mom and dad. It was the club's annual meeting—always well attended—and as I stepped to the podium I looked out over a sea of skeptical faces, the faces of affluent conservatism. I spoke for an hour—about why the environment is so important to the physical and spiritual health of our nation and its people, about how a wholesome environment and a healthy democracy are intertwined, and about the way that President Bush is allowing certain corporations to destroy our country's most central values. I pulled no punches, and I got a standing ovation.

A month before, I got a similar response at the Woman's Club of Richmond, Virginia, where someone boasted that no member had voted for a Democrat since Jefferson Davis. They told me it was the first standing ovation there in 38 years.

Earlier that week I had spoken at an oil-industry association meeting in the Northwest, and I received an equally enthusiastic response.

I got those reactions not because I'm a great speaker (I'm not), but because I talked about the values that define our community and make us proud to be Americans—shared values that are being stolen from us. Those oil executives, Richmond Republicans, and Round Hill Club members have the same aspirations for their children as I have for mine: clean air and water, robust health, beautiful landscapes in which to play and grow and be inspired, and a community that stands for something good and noble.

I want to be very clear here: This book is not about a Democrat attacking a Republican administration. During my two decades as an advocate for the Natural Resources Defense Council, Riverkeeper, and the Waterkeeper Alliance, I've worked hard to be nonpartisan. The fishermen and farmers whom I represent as an attorney run the political spectrum, and I've supported both Democratic and Republican leaders with sound environmental agendas.

Moreover, I don't believe there are Republican or Democratic children. Nor do I think that it benefits our country when the environment becomes the province of one party, and most national environmental leaders agree with me. But today, if you ask those leaders to name the greatest threat to the global environment, the answer wouldn't be overpopulation, or global warming, or sprawl. The nearly unanimous response would be George W. Bush.

You simply can't talk honestly about the environment today without criticizing this president. George W. Bush will go down as the worst environmental president in our nation's history. In a ferocious three-year attack, his administration has launched over 300 major rollbacks of U.S. environmental laws, rollbacks that are weakening the protection of our country's air, water, public lands, and wildlife.

Such attacks, of course, are hardly popular. National polls consistently show that over 80 percent of the American public—with little difference between Republican and Democratic rank and file—want our environmental laws strengthened and strictly enforced. In a March 2003 memo to party leadership, Republican pollster Frank Luntz noted: "The environment is probably the single issue on which Republicans in general and President Bush in particular are most vulnerable." He cautioned that the public is inclined to view Republicans as being "in the pockets of corporate fat cats who rub their hands together and chuckle maniacally as they plot to pollute America for fun and profit." If that view were to take hold, Luntz warned, "not only do we risk losing the swing vote, but our suburban female base could abandon us as well." In essence, he recommended that Republicans don the sheep's clothing of environmental rhetoric while continuing to wolf down our environmental laws.[1]

White House strategists grasped that lesson long before the Luntz memo. The administration has gone to great lengths to keep the president's agenda under wraps, orchestrating the legislative rollbacks almost entirely outside of public scrutiny. It has manipulated and suppressed scientific data, intimidated enforcement officials and other civil servants, and masked its agenda with Orwellian doublespeak. Bush's "Healthy Forests" initiative promotes destructive logging of old-growth forests.

His "Clear Skies" program suggests repealing key provisions of the Clean Air Act. The administration talks about "streamlining" and "reforming" regulations when it means weakening them, and "thinning" when it means logging or clear-cutting. Cloaked in this meticulously crafted language that is designed to deceive the public, the administration—often unwittingly abetted by a toothless and negligent press—intends to effectively eliminate the nation's most important environmental laws by the end of its term.

But this book is ultimately about more than the environment. It's about the corrosive effect of corporate cronyism on free-market capitalism and democracy—core American values that I cherish. There are, of course, good and even exemplary corporations in every sector. Even in the oil business, companies like BP, Shell, and Hess have acted aggressively to deal with global warming and have behaved responsibly toward the environment. But corporations, no matter how well intentioned, should not be running the government.

This administration, however, in its headlong pursuit of private profit and personal power, has sacrificed respect for the law, private property rights, scientific integrity, public health, long-term economic vitality, and commonsense governance on the altar of corporate greed.

Our government has abandoned its duty to safeguard our health and steward our national treasures, eroding not just our land, but our nation's moral authority and capacity to fulfill its historic mission—to create communities that are models for the rest of humankind. After all, we protect nature not (as Rush Limbaugh likes to say) for the sake of the trees and the fishes and the birds, but because it is the infrastructure of our communities. If we want to provide our children with the same opportunities for dignity and enrichment as those our parents gave us, we've got to start by protecting the air, water, wildlife, and landscapes that connect us to our national values and character. It's that simple.

The Bush attack was not entirely unexpected. During his tenure in Texas, George W. Bush had the grimmest environmental record of any governor in the country: the Lone Star State ranked number one in both air and water pollution. In his six years in Austin, Governor Bush championed a short-term, pollution-based prosperity that enriched his political contributors and corporate cronies by lowering the quality of life for everyone else. Now President Bush is doing the same thing to the citizens in the other 49 states.

The present cabinet boasts more CEOs than any in history. Most come from the energy, extractive, and manufacturing sectors that rely on giant subsidies and create the worst pollution. Almost all the top positions at the agencies that protect our environment and oversee our resources have been filled by former lobbyists for the biggest polluters in the very businesses that these ministries oversee. These men and women seem to have entered government service with the express purpose of subverting the agencies they now command. The administration is systematically muzzling, purging, and punishing scientists and other professionals whose work impedes corporate profit taking. The immediate beneficiaries of this corrupt largesse have been the nation's most irresponsible mining, chemical, energy, agribusiness, and automobile companies. The American people have been the losers.

Environmental injury is deficit spending—loading the costs of pollution-based prosperity onto the backs of the next generation. In 2003 the Environmental Protection Agency announced that for the first time since the Clean Water Act was passed 30 years ago, American waterways are getting dirtier. In Lake Erie, painstakingly resurrected by the Clean Water Act, the infamous dead zone is expanding once again. More raw sewage is flowing into our rivers, lakes, and streams as the White House throws out rules designed to end sewer-system overflows. Bush's policies promote greater use of dangerous pesticides, deadly chemicals, and greenhouse gases, and encourage the filling of wetlands and streams. The administration has removed protections from millions of acres of public lands and wetlands and thousands of miles of creeks, rivers, and coastal areas.

I am angry both as a citizen and a father. Three of my sons have asthma, and on bad-air days I watch them struggle to breathe. And they're comparatively lucky: One in four African American children in New York City shares this affliction, and many lack the insurance and high-quality health care that keep my sons alive and active.[2] Sadly, too, few children today can enjoy that quintessential American experience, going fishing with Dad and eating their catch. Most bodies of water in New York—and all freshwater bodies in 17 other states—are so tainted with mercury that one cannot eat the fish with any regularity. Forty-five states advise the public against regular consumption of at least some local fish due to mercury contamination.

I often take my children to hike, fish, and canoe in the nearby Adirondack Mountains, the oldest protected wilderness on Earth. Since the area was declared "forever wild" in 1885, generations of Americans might reasonably have expected to enjoy its unspoiled rivers and streams. But 500 lakes and pools (out of 2,800) in the Adirondacks have now been rendered sterile by acid rain.

The mercury and the pollutants that cause acid rain and

provoke most asthma attacks come mainly from the smokestacks of a handful of outmoded coal-burning power plants. These discharges are illegal under the Clean Air Act. But President Bush recently sheltered these plants from civil and criminal prosecution, and then excused them from complying with the act. Amazingly, his administration is instead relying on a cleanup schedule written by polluters for polluters that will leave the United States with contaminated air, poisoned water and fish, and sickened children for generations. The energy industry, by the way, gave $48 million to President Bush and his party during the 2000 campaign, and have ponied up another $58 million since. They are now reaping billions of dollars in regulatory relief. But generations of Americans will pay that campaign debt with poor health and diminished lives

Furthermore, the addiction to fossil fuels so encouraged by White House policies has squandered our Treasury, entangled us in foreign wars, diminished our international prestige, made us a target for terrorist attacks, and increased our reliance on petty Middle Eastern dictators who despise democracy and are hated by their own people.

Several of my own lawsuits have been derailed by George W. Bush and friends. As he began his presidency, I was involved in litigation against the factory-pork industry, which is one of the largest sources of air and water pollution in the United States. Industrial farms illegally dump millions of tons of untreated fecal and toxic waste onto the land and into the air and water. Factory farms have contaminated hundreds of miles of waterways, put tens of thousands of family farmers and fishermen out of work, killed billions of fish, sickened consumers, and subjected millions of farm animals to unspeakable cruelty.

On behalf of several farm and fishing groups, we sued one of the largest hog conglomerates, Smithfield Foods, and won a decision that suggested that almost all large factory farms were violating the Clean Water Act. Then the Bush administration ordered the EPA to halt its own Clean Air Act investigations and weakened the Clean Water rules, neutralizing my lawsuits and allowing the industry to continue polluting indefinitely.

For 20 years, as attorney for Hudson Riverkeeper, I've worked with commercial and recreational fishermen and riverfront communities to force General Electric to clean up the polychlorinated biphenyls that the company has dumped into the river for decades. These PCBs have put hundreds of commercial fishermen out of work, dried up the river's barge traffic (because the shipping channels are too toxic to dredge), contaminated waterfront towns, and infected virtually every person in the Hudson Valley. (My own PCB levels are double the national average!) In February 2002, we finally forced the EPA to sign the long-awaited order requiring the company to dredge the river and recover its PCBs. But our celebration was short-lived.

In October 2003, after President Bush failed to renew an environmental tax on oil and chemical companies, Superfund went bankrupt. With no money in the fund, the EPA has lost its leverage to force General Electric to act. The EPA's principal leverage over recalcitrant polluters was Superfund's treble damages provision, which allows the agency to use the fund to clean up the site and then charge the polluter three times its costs. "I do not believe that the Hudson will ever be cleaned up by General Electric, except under threat of the treble damages provision, and that no longer exists," says Janet MacGillivray, the EPA's former assistant regional counsel. "The company has already avoided responsibility for thirty years. Without that leverage, General Electric could conceivably litigate this case for decades."[3] Without that cleanup, the Hudson, according to the best federal science, will be polluted for my lifetime and that of my children, its fish unsafe to eat for the next century. Thanks to President Bush's decision, one out of every four Americans lives within a few miles of a Superfund site that may never be cleaned up.

The fishermen, farmers, and other working people whom I represent are by and large traditional Republicans who live by Teddy Roosevelt's precept: "The nation behaves well if it treats the natural resources as assets which it must turn over to the next generation increased, and not impaired, in value."[4] Without exception, these people see the current administration as the greatest threat not just to their livelihoods but to their values, their sense of community, and their idea of what it means to be American. Why, they ask, is the president allowing coal, oil, power, chemical, and automotive companies to fix the game?

1

The Mess in Texas

As you fly over the Houston Ship Channel at twilight, thousands of flares seem to ignite in the approaching darkness. Smokestacks from more than a hundred massive chemical factories, oil refineries, and power plants have suddenly become steel towers of light and fire. From the air, it's not hard to understand why some call this area the "golden triangle." This concentration of industry, which includes a 3,000-acre ExxonMobil facility—the planet's largest oil refinery—generates enough wealth for its owners to make the Texas economy bigger than the gross domestic product of most nations.[1]

It is a different scene on the ground. There the twilight flares rumble, the ground shakes, the air hisses. Plumes of black smoke belch upward and acrid odors permeate the atmosphere. The smell of money, some call it. But from this earthly vantage point—especially for low-income residents living downwind in eastern Harris County—it is less a golden triangle than a scene out of Dante's *Inferno.*

The ubiquitous highway signs warning "Don't Mess With Texas," haven't deterred the state's polluters one bit. Here are some basic facts about the Lone Star State: According to the Texas Commission on Environmental Quality, fully one-quarter of Texas's streams and rivers are so polluted that they do not meet standards set for recreational use.[2] Half of the state's 20 million people reside in areas where the smog pollution surpasses federal limits.[3] In 1999, Houston overtook Los Angeles as America's smoggiest city. Texas also ranked first in toxic releases to the environment, first in total toxic air emissions from industrial facilities, first in toxic chemical accidents, and first in cancer-causing pollution.[4] Also in 1999, 15 of the nation's 30 highest smog readings were all taken in Texas.[5] Every major urban area—Houston, Dallas, San Antonio, Austin, El Paso, and Longview—either failed to meet the EPA's minimum air quality standards, or was on the verge of failing.[6]

"The level of damage to human health is extraordinary," says Tom Smith, director of the Texas office of Public Citizen, a consumer advocacy organization. He cites a recent mayoral study estimating annual pollution-related health care costs of between $2.9 and $3.1 billion in the Houston metropolitan area alone.[7] Air pollution kills an estimated 435 people a year in the city.[8] "We lead the nation in childhood asthma," says Lanell Anderson, a resident of Clear Lake, a town south of Houston that's surrounded by chemical plants. "We lead the nation in childhood cancer. . . . Our cup runneth over."[9]

Texas has long been one of the most polluted states in the country, but rather than remedy the situation, George W. Bush set out to destroy virtually all attempts to clean up the state's tainted air, water, and land. During his six-year reign as governor, from 1994 to 2000, Texas dropped to number 49 in spending on the environment.[10] Under his watch, Texas had the worst pollution record in the United States. It sent the most toxic chemicals and carcinogens into the air. It had the highest emissions of carbon dioxide (CO_2), accounting for at least 10 percent of the national total. It had the most chemical spills and Clean Water Act violations, and produced the largest volume of hazardous waste.[11] As *New York Times* columnist Bob Herbert put it shortly before Bush received the Republican nomination in 2000, "Mr. Bush's relationship to the environment is roughly that of a doctor to a patient—when the doctor's name is Kevorkian."[12]

The anti-environment agenda of today's White House was honed and perfected during Bush's gubernatorial years. It was in Texas that he developed the tactics and policies that guide his autocratic leadership today: closed-door meetings with industry insiders who are among his biggest campaign contributors; reliance on pseudo-scientific "studies" by right-wing think tanks; emasculation of regulations that cut into industry profits; citizens muzzled in debates that affect their communities.

Soon after becoming governor, Bush declared tort reform an "emergency issue" and appointed judges who made it all but impossible for Texans to bring class action lawsuits against polluters. In 1995 he pushed through the Private Real Property Rights Preservation Act, a radical "takings" bill that would make taxpayers pay polluters' cost of complying with pollution laws. According to this view, corporations should be able to do what they want with their private property; if the state cuts into their profits by forcing them to adopt pollution-control measures, the state (i.e., the public) should pay. This perverse doctrine reverses a millennium of western property law that holds that owners can use their property as they please, but never in a way that diminishes their neighbors' property or the public trust properties like air and water. Leading the charge for this radical new approach was right-wing private-property advocate Marshall Kuykendall, who complained at a public forum that the last time the federal government took our property without compensation is "when Lincoln freed the slaves."[13]

In another foreshadowing of his presidency, Bush installed a pro-industry troika to run the state's environmental agency, the Texas Natural Resources Conservation Commission. Bush selected Barry McBee, a lawyer with a host of oil-industry clients, to chair the TNRCC. At his previous position at the Texas Department of Agriculture, farm labor and environmental groups accused McBee of helping to dismantle a program that kept farmworkers out of fields that were still "hot" after pesticide applications. The second appointee was Ralph Marquez, a former Monsanto executive and lobbyist for the Texas Chemical Council. Marquez quashed a plan to issue health warnings to Houston residents on high-smog days and later testified before a congressional committee that ozone "is a relatively benign pollutant."[14] Bush's third appointment was a cattleman named John Baker, former official of the Texas Farm Bureau, a sworn enemy of pesticide regulations.[15]

The new TNRCC came to be known by the moniker "Train Wreck." Until this new regime was in place, all Texas citizens had the right to challenge pollution permits required by companies for their waste disposal. This right is one of the few recourses that regular folks have to protect their health, homes, and communities from the ravages of pollution. The new TNRCC soon eliminated this policy, as well as the longstanding practice of making surprise inspections of industrial plants. It discovered loopholes in all kinds of federal and state environmental regulations. On Halloween 1995, for example, the TNRCC announced Texas's plan to revise the Ozone National Ambient Air Quality Standards, an EPA directive that requires states to monitor for unsafe levels of ozone. The TNRCC decided it would mathematically average ozone pollution across large areas, in hopes that, in the words of Neil Carman, a former agency staffer, it could make "exceedances disappear by massaging the high numbers." Carman is now Clean Air Director for the Lone Star chapter of the Sierra Club.[16]

Slashing the TNRCC's budget by 20 percent, Bush ensured that the commission couldn't possibly fulfill its duty as the state's environmental watchdog. Texas virtually ceased monitoring water quality after Bush's election, for example, despite the fact that Texas had far more facilities discharging into waterways than any other state.[17] The Environmental Working Group, a national nonprofit research organization, reported that Texas also had the worst record in the country for inspecting companies that violated the Clean Water Act.[18] Indeed, so little money was spent on protecting waterways that "almost nothing is known about the quality of 25,000 out of 40,000 miles of the state's permanent rivers and streams," according to the *Texas Environmental Almanac* in 1995. Even when the TNRCC did know of toxic water, it often failed to disclose its findings to the public. When the commission learned of high mercury levels in the Rio Grande River near Laredo, for example, it refused to inform residents.[19] In 1999, the Natural Resources Defense Council named Texas as one of six "beach bum" states for a second consecutive year—because the state had no monitoring system designed to alert swimmers to potential pollution-related health risks.[20]

But it is Governor Bush's record on air pollution that is most appalling. When the Texas Clean Air Act of 1971 became law, more than 1,000 industrial facilities were "grandfathered," or exempted from the new pollution regulations. The idea was that these grandfathered plants would eventually either modernize or become obsolete and close down. This was wishful thinking at best: In reality, companies that didn't have to spend money on pollution control had a competitive edge over their regulated competitors.[21] And with little incentive to modernize, they didn't. While their competitors had to apply for a permit to pollute, running the gauntlet of public comment and government scrutiny, grandfathered companies just kept their outdated plants up and running.

These grandfathered polluters now create havoc for communities all over Texas. For example, some 30 miles from Dallas is the town of Midlothian, known as the "Cement Capital of America." The largest operator is the TXI Corporation, whose emission control systems date back to 1972. The plant is powered by a generator that burns hazardous waste trucked in by other companies, a double-your-money idea that eliminates the need for natural gas to fire TXI's kilns. The company's own testing has revealed smokestacks belching out carcinogens at levels far in excess of EPA standards.[22] According to a 1997 report in the *Dallas Observer,* "scientists don't even have names for some of the substances coming out of TXI's stacks."[23] Midlothian residents have long complained of a variety of health problems, and a 1996 report by the Texas Department of Health noted that Down's syndrome is unusually prevalent in babies born in the area.[24]

Fully one-third of the state's air pollution—903,800 tons a year by the end of the 1990s—was issuing from these grandfathered smokestacks.[25] These plants emit as much nitrogen oxide as 18 million cars.[26] One company, Alcoa, Inc., North America's largest aluminum smelter, was responsible for more than 100,000 tons of toxic emissions at one of its plants in 1997. Neighboring Milam County residents maintain that the air around Alcoa's smelter is so acidic that it eats the galvanized coating off barbed-wire fences. Neil Carman of the Sierra Club compares the situation to getting caught driving without a license, but happily finding that speed limits don't apply to you: "You just say, 'Well, I'm grandfathered, officer,' and the reply is, 'Have a good day, just don't kill too many people.'"[27]

By the 1990s, the Clinton administration's EPA had these Texas polluters in its crosshairs. This put Bush in a tight spot. In order to get the EPA off his back, he needed to put some kind of state regulations in place. Trouble was, many of the top polluters were his prime financial backers. Between 1993 and 1998 Alcoa, Exxon, Shell, Amoco, Enron, Dow Chemical, and others poured $1.5 million into the Bush campaign coffers.[28]

After privately conferring with these corporate backers, Bush pushed through two landmark laws, the 1995 Texas Audit Privilege and Immunity Law, and the 1999 Voluntary Emissions Reduction Permit Program. Rather like the fox guarding the henhouse, the laws allow companies to monitor their own pollution, report their violations to the government, and promise to clean up. There would be no fines, no public disclosure, no government follow-up. Imagine a world where criminals could stay out of jail simply by confessing their crime to a state agency and promising to do better in the future. That's the law Bush came up with to cope with some of the most polluted air on Earth.

The outcome of these two laws does not bode well for the rest of America. In the summer of 2000, in a news item that received little national attention, the University of Texas released the results of its $20 million Texas Air Quality Study. The study revealed that Houston's industries had been vastly underreporting the magnitude of their emissions. Scientists found it hard to believe the instrument readings from aircraft measuring industrial plumes. The various chemicals were 6 to 15 times higher than what the Houston Ship Channel's factories had been reporting to state and federal agencies. Some chemicals were at levels 100 times higher.[29] In 2001, the TNRCC's draft report to the Texas legislature conceded that Governor Bush's plan had so far resulted in zero reductions in air pollution.

This was hardly news to most Texans. Back in October 1999, in the Houston suburb of Deer Park, high school athletes exercising outside experienced severe coughing fits and difficulty breathing in the midst of one of the worst smog episodes on record. Furious parents demanded that the state notify schools to protect students from outdoor exertion during severe air pollution episodes.[30] According to the Texas chapter of Public Employees for Environmental Responsibility (PEER), more than 290 schools were located within a two-mile radius of a "grandfathered" polluter.[31] So much for No Child Left Behind.

Yet Bush no doubt considers his air pollution laws wildly successful. The CEOs of at least nine of Texas's grandfathered polluters ponied up a minimum $100,000 pledge to become "Bush Pioneers" in support of their pal's bid for the White House. Alcoa's law firm, Vinson & Elkins, was also a top contributor. So was another litigator, Baker & Botts, the law firm of James Baker, who would guide Bush's legal fight in Florida after the disputed 2000 election. The Baker & Botts client list included eight of the grandfathered polluters.[32] "Pollution policy in Texas has become a cash-and-carry operation," observes Erin Rogers, coordinator of Texas's PEER office. "If you have the cash, you can carry on as you like."[33]

When George W. Bush became president, he brought his environment-for-sale agenda with him, along with many of his Texas cronies. At the top of the list, of course, was Vice President Dick Cheney, CEO of the world's second-largest oil-drilling services company, Texas-based Halliburton. The Energy Department's transition team, also headed by Cheney, included three Bush Pioneers from Texas, one of whom was Enron's former chairman and CEO Kenneth Lay. To run the Commerce Department, Bush tapped another oilman from Midland, Don Evans. Karl Rove, then a right-wing Texas political consultant with long-term ties to the Bush family, became his chief political adviser. Tom DeLay was already on Capitol Hill. DeLay, the onetime Houston pest exterminator turned House whip, once referred to the EPA as "the Gestapo of government."[34] This core group would in turn bring along their own friends of industry, such as Paul O'Neill, former CEO of Alcoa, whom Bush tapped to run the U.S. Treasury Department.

A Charge to Keep, George W. Bush's 253-page book on his pre-presidential life and times, contains a single sentence on air and water pollution.[35] That's one more sentence than he devoted to the environment in his 2004 State of the Union speech. But as we'll soon see, the environment has remained very much on his mind—with the lessons he learned in Texas serving as his guide.

2

Back to the Dark Ages

The fact is, it's always been illegal to pollute. The protection of the shared environment has been one of government's most fundamental roles since constitutions were devised. Ancient Rome's Code of Justinian guaranteed to all citizens the use of the "public trust," or commons—those shared resources that cannot be reduced to private property, including the air, flowing water, public lands, wandering animals, fisheries, wetlands, and aquifers.

Throughout Western history the first acts of tyrants have invariably included efforts to deliver the public trust assets into private hands. When Roman law broke down in Europe during the Dark Ages, feudal kings began to privatize the commons. The legendary outlaw Robin Hood became a potent symbol of defiance against King John's efforts to reserve England's deer and wildlife for the privileged classes. When King John attempted to sell off the country's fisheries and to erect navigational tolls on the Thames, the public rose up and confronted him at

Runnymede in 1215, forcing him to sign the Magna Carta, which includes provisions guaranteeing the rights of free access to fisheries and waters. In thirteenth-century England it was a capital offense to burn coal in London, and violators were executed for the crime.[1]

These public trust rights to unspoiled air, water, and wildlife passed to the people of the United States following the American Revolution. Everyone had the right to use the commons, but never so as to injure its use and enjoyment by others. Until 1870, a factory releasing even small amounts of smoke onto public or private property was operating illegally, and courts had no option but to forbid the activity. As late as 1913, the U.S. Supreme Court declared that it was "inconceivable that public trust assets could slip into private hands."

Beginning in the Gilded Age, however, when the corporate robber barons captured the political and judicial systems, those rights were stolen from the American people. Judges and legislators, who were either corrupted or convinced of the merits of unfettered industrial development, began to dilute the public right to be free from pollution. As the Industrial Revolution transitioned into the postwar industrial boom, Americans found themselves paying a high price for the resulting pollution. The wake-up call came in the late 1960s, when Lake Erie was declared dead, Cleveland's Cuyahoga River caught fire, and radioactive strontium 90 was found in mothers' breast milk across North America and in even the most remote corners of the globe.

As a boy I was conscious of these events. I wasn't allowed to swim in the Hudson, the Potomac, or the Charles Rivers. I remember dusting our home daily for soot from the black smoke that billowed from stacks around Washington, D.C. I was aware that thousands of Americans died each year during smog events. I never saw herons, ospreys, and bald eagles. Pesticides had mostly extinguished their mid-Atlantic populations and were doing the same to our songbirds. But I do remember the Eastern Anatum peregrine falcons that used to nest on the old post office building on Pennsylvania Avenue in Washington. They were the most beautiful of all the peregrine subspecies—salmon pink with a white coverlet over their cere—and they could reach speeds of over 240 miles per hour. They were the fastest birds in the world. As a young falconer, I loved to watch their vertical stoops to pick pigeons from the air in front of the White House. It's a sight my children will never see, because that bird went extinct in 1963, the same year my uncle Jack died. They were poisoned out of existence by DDT—a million years of evolution sacrificed to ignorance and greed in the blink of an eye.

On Earth Day in 1970, the accumulation of such insults drove 20 million Americans to the streets in the largest public demonstrations in U.S. history. Motivated by that stunning display of grassroots power, Republicans and Democrats, working together, created the Environmental Protection Agency and passed 28 major laws over the next 10 years to protect our air, water, endangered species, wetlands, food safety, and public lands.

Statutes like the Clean Air Act, the Clean Water Act, and the Endangered Species Act were designed to both protect the environment and strengthen our democracy. They made government and industry more transparent on the local level. Powerful corporate polluters would finally be held accountable: Those planning to use the commons would have to deal with environmental impacts and submit to public hearings; citizens were given the power to prosecute environmental crimes. Even the most vulnerable Americans could now participate in the dialogue that determines the destinies of their communities. The passage of these statutes marked the return of ancient rights recognized since the dawn of civilization.

The victory was short-lived. Earth Day 1970 caught polluters off guard, but over the next 30 years they mounted an increasingly sophisticated and aggressive counterattack to undermine these laws. The environmental reversals of the Bush administration are the triumphant outcome of their three-decade campaign.

Their success has been largely the result of an unholy marriage between polluting industries and the radical right—an alliance conceived by Colorado brewer Joseph Coors. In 1976, Coors, owner of one of Colorado's biggest polluters, founded the Mountain States Legal Foundation (MSLF) to challenge environmental laws.[2] The MSLF mimicked the work of public interest organizations like the Natural Resources Defense Council, which has been fighting frontline legal battles for the environmental movement since 1970. Funded by multinational polluters such as Phillips Petroleum, Exxon, Texaco, Amoco, Shell, Ford Motor Company, and Chevron, the MSLF filed suits intended to block efforts by environmentalists, unions, minorities, and handicapped Americans that might cut into corporate profit taking.[3]

Coors also founded the right-wing Heritage Foundation, which has provided the philosophical underpinning of the anti-environmental movement.[4] The Heritage Foundation's function is to produce short, concise policy analyses of fast-breaking issues. These simple position papers go out to thousands of news directors and journalists, congressional offices, public officials, and hate-radio jocks. Through clever invocations of patriotism, Christianity, and laissez-faire capitalism,

Heritage offers pithy philosophical justifications for national policies that promote the narrow interests of a wealthy few.

From its inception, the Heritage Foundation urged its followers "to strangle the environmental movement," which it declared "the greatest single threat to the American economy," dismissing global warming, acid rain, and other environmental crises as "henny pennyism."[5] Its prominence as the leading voice for pollution-based prosperity helped it attract enormous donations from the automobile, coal, oil, and chemical companies. Heritage became a top beneficiary of five right-wing foundations established by major corporate polluters: the John M. Olin Foundation, funded by a leading manufacturer of ammunition and toxic chemicals; the Sarah Scaife Foundation, funded by the Mellon industrial, oil, and banking fortune and run by the arch-conservative *American Spectator* financier Richard Mellon Scaife; the Castle Rock Foundation, funded by Coors; the Charles G. Koch Charitable Foundation, headed up by Charles G. Koch, chairman of Koch Industries, Inc., the largest independent oil company in the United States and one of the biggest polluters in Texas; and the Bradley Foundation, funded by the electronics fortune.[6]

This "Gang of Five" has helped keep Heritage in business, providing it with close to $40 million since 1985. It and a gaggle of corporate polluters, including ExxonMobil, Chevron, and DuPont, continue to provide a large portion of Heritage's $34 million annual budget.[7] They have also helped to create an array of Heritage look-alikes: the Competitive Enterprise Institute, the American Enterprise Institute, the Reason Foundation, the Federalist Society, the Marshall Institute, and the secretive Mercatus Center, all of which get additional funding from the filthiest players in the oil, chemical, timber, mining, and agribusiness sectors.

The purpose of these so-called think tanks is to run interference for their corporate funders and provide a counterweight to the public interest groups shining a spotlight on their antisocial activities. They fight to exempt industry from toxic waste laws, to open wilderness areas and national parks to clear-cutting and mineral extraction, and to lift wetland protections. At the same time, they provide industry and politicians with the cover they need to pretend that there is a genuine debate over the objectives of the environmental movement.

Around 1980, Coors and Co. spearheaded the Sagebrush Rebellion, a coalition of industrial heavyweights and right-wing ideologues that set out to turn their think-tank policies into political power. When candidate Ronald Reagan declared, "I am a Sagebrush rebel," the big polluters were elated. Reagan's victory gave the Heritage Foundation and the MSLF a national arena for their radical agenda. Heritage became known as Reagan's "shadow government,"[8] and its 2,000-page manifesto, *Mandate for Change,* became the blueprint for his administration. Joe Coors headed the cabal of right-wing millionaires that formed Reagan's kitchen cabinet, which set up shop in the Executive Office Building directly across from the White House.[9]

Coors handpicked his Colorado associate Anne Gorsuch to administer the EPA. He chose her husband-to-be, Robert Burford, a subsidy-dependent cattle baron who had vowed to destroy the Bureau of Land Management, to head up that very agency.[10] Coors chose James Watt, president of the MSLF, as the secretary of the Department of the Interior. Watt was a proponent of "dominion theology," an authoritarian Christian heresy that advocates man's duty to "subdue" nature. His deep faith in laissez-faire capitalism and apocalyptic Christianity led Secretary Watt to set about dismantling his department and distributing its assets, selling off public lands and water and mineral rights at what the General Accounting Office called "fire-sale prices."[11] During a Senate hearing, Mr. Watt cited the approaching Apocalypse to explain why he was giving away America's sacred places rather than preserving them for future generations: "I do not know how many future generations we can count on before the Lord returns," he explained.[12] Watt believed that environmentalism was a plot to delay energy development and "weaken America," and dismissed environmentalists as "a left-wing cult which seeks to bring down the type of government I believe in."[13]

Meanwhile, Anne Gorsuch enthusiastically gutted the EPA's budget by 30 percent, crippling the agency's ability to write regulations or enforce the law. She purposely destroyed the Superfund program at its birth, turning it into a welfare program for industry lawyers. She appointed lobbyists fresh from their hitches with paper, asbestos, chemical, and oil companies to run each of the principal agency departments.[14] Her chief of staff was a timber-industry lawyer; her enforcement chief was from Exxon.[15]

These attacks on the environment precipitated a public revolt. By October 1981, 1.1 million Americans had signed a petition demanding Watt's removal.[16] After being forced out of office, he was indicted on 25 felony counts of perjury, unlawful concealment, and obstruction of justice.[17] Gorsuch and 23 of her cronies were forced to resign following a congressional investigation of sweetheart deals with polluters, including Coors.[18] Her first deputy, Rita LaVelle, was jailed for perjury and obstruction of justice.[19]

The indictments and resignations put a temporary damper on the Sagebrush Rebels, but they quickly regrouped. During an August 1988 conference at the Nugget Hotel in Reno, Nevada, they adopted a new moniker, "Wise Use," cynically chosen to imply a thoughtful approach to the environment.[20] As Wise Use founder and timber industry flack Ron Arnold put it, "Our goal is to destroy, to eradicate the environmental movement. We want you to be able to exploit the environment for private gain, absolutely."[21]

By using its vast financial resources, Arnold urged, industry could control its destiny in the courts and Congress, actually dictating favorable legislation through its grassroots arms "guided by signals from . . . industry's professional lobbyists."

Following the meeting in Reno, hundreds of small Wise Use groups began to pop up across the United States. Usually they focused on some local issue—development, lumber, mining, or grazing. While a few had genuine grassroots support, most were industry front groups organized by public relations consultants who specialize in "greenwashing"—deceiving the public on environmental issues—now a half-billion-dollar industry. These groups adopted environmentally friendly names to mask their purpose. The Citizens For the Environment (CFE), for example, has no citizen membership and gets its support from a long list of corporate sponsors who use the organization to lobby against the Clean Air Act and other environmental regulations. The Environmental Conservation Organization is a front group for land developers and other businesses opposed to wetlands regulations. The Evergreen Foundation is a timber-industry mouthpiece that promotes the idea that clear-cut logging is beneficial to the environment. Citizens for Sensible Control of Acid Rain is a front for the oil and electric industries that is opposed to *all* controls of acid rain.

Though small in number, these phony grassroots, or "Astro-

Turf," groups had a disproportionate impact due to their access to nearly unlimited industry resources, the right wing's network of think tanks, and the voices of sympathetic hate-radio jocks like Bob Grant and Rush Limbaugh.

But the most important vector for hammering the Wise Use agenda into the Republican Party's platform was the Christian right.

From the start, the Wise Use movement was closely linked to a handful of powerful, authoritarian, right-wing Christian leaders. For instance, the convicted tax felon Reverend Sun Myung Moon and his Unification Church, which owns the right-wing *Washington Times,* underwrote the costs of the Reno conference and provided seed money for dozens of Wise Use groups. Ron Arnold is head of the Washington State chapter of the American Freedom Coalition, the political arm of the Unification Church.[22]

But when the Wise Use allies hooked up with Pat Robertson's Christian Coalition, they hit a home run. Robertson's special contribution to right-wing theology was to substitute environmentalists for communists as the new threat to democracy and Christianity. In his 1991 best-seller, *The New World Order,* he vilifies the federal government as an alien nation waging war on the family and disarming America through gun control laws. Environmentalists are the evil priests of a new paganism that will become the official state religion of the New World Order. These ravings would hardly be worth mentioning had they not played such an important role in forming the ideological underpinnings of the anti-environmental movement and fueling the zealotry of its followers, which now include many high-ranking officials in the Bush White House and in Congress.[23] Robertson's aggressive anti-environmental proselytizing has opened the door for Christian extremists, militia aficionados, and white supremacists from the fringe who enthusiastically adopted the issue for their own purposes.[24]

But Robertson has also helped make anti-environmentalism acceptable within the ranks of the fundamentalist clergy and the mainstream of the Republican Party. Beginning in 1991, Robertson and the Christian Coalition's then–executive director Ralph Reed, now an official with the Bush campaign, put their media and organizational clout at the disposal of the Wise Use agenda. While Robertson made anti-environmentalism a principal theme on his Christian Broadcasting Network talk shows, news hours, and documentaries, Reed gave seminars to corporate public relations executives, coaching them on how to use electronic technologies and grassroots organizing to foil environmentalists who interfere with polluter profits.[25]

Robertson's brand of paranoia has always had a place in American politics, from the populist movement to Father Coughlin to the John Birch Society. But until 1994, it was never able to achieve the kind of power that only comes from money. Suddenly the extractive and chemical industries, their profits threatened by new environmental laws, saw the Christian right's anti-environmentalism as a ticket to continued prosperity and donated the funds that sent its candidates to Congress.

In 1994, industry's greenwashing and its years of investment in political organizations, front groups, think tanks, and phony science paid off in the most pro-pollution Congress in our nation's history. Wise Use helped propel Newt Gingrich to the Speaker's chair of the U.S. Congress, where he began a dangerous and partially successful effort to enact his anti-environmental manifesto, Contract With America. Gingrich's *consigliore* was Congressman Tom DeLay, the former bug exterminator who was determined to rid the world of pesky pesticide regulations and to promote a "biblical worldview."[26] DeLay considers DDT "safe as aspirin"[27] and the Endangered Species Act the greatest threat to Texas after illegal aliens.[28] He attributed the Columbine massacre to the teaching of evolution in schools.[29] In January 1995, Congressman DeLay invited a group of 350 lobbyists representing some of the nation's biggest polluters to collaborate in drafting legislation that would dismantle federal health, safety, and environmental laws.[30]

The followers of Gingrich and DeLay had learned from the James Watt debacle that they had to conceal their radical agenda. Under the tutelage of Republican pollster Frank Luntz—who drafted the Contract With America—they attended tree-planting ceremonies and greenwashed their language to attack big government and excessive regulations and to laud property rights and free markets.

Carefully eschewing public debate, they mounted a stealth attack on America's environmental laws. Rather than a frontal assault against popular statutes like the Endangered Species Act, the Clean Water Act, and the Clean Air Act, they tried to undermine these laws by silently attaching riders to must-pass budget bills or by promoting "supermandates" with seductive names claiming to support "regulatory reform" or "property rights" and to oppose "unfunded mandates." Each was designed to eviscerate whole bodies of environmental law without debate.

But the public got wise. The NRDC, the Sierra Club, and the U.S. Public Interest Research Group (U.S. PIRG) helped direct more than a million letters to Congress, and moderate Republicans teamed with the Clinton administration to block the worst of it. When President Clinton shut down the government in December 1995 rather than pass a budget bill spangled with anti-environmental riders, the American public turned against Gingrich, DeLay, and their accomplices. By the end of that month, even conservatives publicly disavowed the attack. Tom DeLay openly admitted that the Republican leadership had miscalculated. "I'll be real straight with you," he told the *Wall Street Journal.* "We have lost the debate on the environment. I can count votes."[31]

It was only a temporary setback. With seemingly unlimited industry money, the Wise Use movement hardly paused before mounting its most audacious effort yet: installing George W. Bush in the White House. When Bush picked Dick Cheney as his running mate, he all but guaranteed support from the key interests that created Wise Use—oil, coal, mining, timber, chemical, pharmaceutical companies, and agribusiness—and from the Christian right.

As a congressman from Wyoming during the 1980s, "Cheney was the go-to guy for oil and for the Wise Use people," says former Montana congressman Pat Williams, who was elected to the House the same year as Cheney. In 1988, Cheney orchestrated the first veto of a wilderness bill in American history by then-president Ronald Reagan. According to Williams, the bill's sponsor, that veto "served as a rallying point for Wise Use people all up and down the Rocky Mountains. The industry and Wise Use coalition are still fueled by that success."[32] From 1992 until he assumed his current office, Cheney served as a "distinguished adviser" to the Center for the Defense of Free Enterprise, which was the launching pad for the Wise Use movement.[33] The CDFE's executive vice president is Ron Arnold, the movement's founder. During this time Cheney was also on the board of the National Legal Center for the Public Interest, a Wise Use think tank.[34]

Bush's victory was the crowning achievement of the Wise Use coalition, and he wasted no time rewarding his benefactors. The assault on the environment began on Inauguration Day, when President Bush's chief of staff and former General Motors lobbyist Andrew Card froze all pending Clinton regulations. Recognizing that policy is personnel, the Bush team installed Wise Use sympathizers in key administration posts. First prize went to Wise Use radical Gale Norton, who got the top post at the Department of the Interior. Her second in command is J. Steven Griles, a notorious former lobbyist for the mining industry. The head of the Department of Agriculture's Forest Service is Mark Rey, a former timber-industry lobbyist. At the Justice Department, the assistant attorney general for Environment and Natural Resources is former mining-industry lobbyist and Wise Use leader Thomas Sansonetti. Until last year, the EPA's deputy administrator was Linda Fisher, a former lobbyist for Monsanto, and Superfund was run by Marianne Horinko, a lobbyist and consultant to polluters, including the Koch Petroleum Group and Koch Industries.[35] The assistant administrator at Air and Radiation is Jeffrey Holmstead, who had been a lobbyist for the utility industry and a leader of a Wise Use industry front group. The director of the White House Council on Environmental Quality, James Connaughton, was a lawyer for asbestos polluters.

Using the stealth tactics originally devised by Gingrich and company, Bush's dream team, in a coordinated effort to implement the Wise Use agenda, has enlisted every federal agency that oversees environmental programs. They have given quick permit approvals and doled out waivers that exempt campaign contributors and polluters from rules or regulations. They have critically reduced funding for implementing environmental laws—for example, defunding the program that lists new endangered or threatened species. They have reinterpreted long-standing policies to limit government authority and to facilitate polluter projects. In May 2004, for instance, the Commerce Department, which oversees U.S. fisheries policy, reinterpreted the Endangered Species Act to allow farm-bred fish to be counted in order to remove salmon from the Endangered Species List and to rid logging companies of restrictions designed to protect wild stocks.

Predictably, the administration has also put the brakes on new rules to protect the environment. In its first three years, Bush's EPA completed just three major rules—two were required under court order and the other rolled back restrictions on power-plant emissions. Compare this to over 30 completed by the Clinton administration and 21 by the former president Bush's administration in their first three years.[36]

As might be expected from a government determined to promote a wildly unpopular agenda, the Bush administration is obsessed with secrecy. In September 2003, the White House moved to weaken the National Environmental Policy Act (NEPA)—the first and most important environmental law—which requires environmental analysis and public participation in major regulatory decisions. And in June 2003 the U.S. Forest Service (USFS) expanded its use of so-called categorical exclusions, which allows the agency to avoid environmental analysis and public scrutiny. In my own experience, the White House has made it dramatically more difficult to obtain material under the Freedom of Information Act or to have unguarded conversations with government officials than at any time in my 20 years as an advocate.

Most galling is the fact that the very agencies entrusted to protect Americans from polluters have simply stopped enforcing the law. Penalties imposed for environmental violations have plummeted under Bush, who has pushed to eliminate 210 enforcement positions.[37] Violation notices have fallen 58 percent, and administrative fines have dropped 28 percent.[38]

In 2001, the White House instructed the EPA's Office of Enforcement and Compliance to stop filing new cases against giant factory farms without approval from upper-echelon political appointees in the EPA.[39] These meat factories release deadly gases such as sulfide, ammonia, methyl mercaptan, methyl sulfides, particulate matter, and airborne animal allergens that cause a range of illnesses, including severe respiratory problems, gastrointestinal diseases, eye infections, nosebleeds, nausea, miscarriage, and psychological problems.[40] Many of these emissions are illegal, and prior to Bush taking office the EPA had been prosecuting these companies under the Clean Air Act and Superfund.

But the Bush administration's order now leaves communities exposed to foul air. "The EPA is no longer a public health agency," says its former chief prosecutor Eric Schaeffer. "It's become a country club for America's polluters."[41]

President Bush is not just letting the multinational meat barons off the legal hook. The White House dropped dozens of Clinton-era prosecutions against its big oil and big coal contributors. One example is the 97-count felony indictment the Clinton administration brought in September 2000 against Koch Industries, the country's largest privately held oil company, owned by Wise Use funder and Bush megadonor Charles Koch. The government said Koch knowingly discharged 90 metric tons of carcinogenic benzene at a refinery in Corpus Christi, Texas, and concealed the releases from federal regulators. The charges could have brought fines of up to $352 million, but after Bush's Department of Justice took over, the counts were reduced from 97 to 9, and the case was settled for

$20 million.[42] Bush has turned down repeated requests by consumer groups for the Federal Trade Commission to investigate price gouging by oil and gas companies, despite a March 2001 FTC finding that companies hoarded gasoline to drive up prices and boost profits, costing consumers billions of dollars.[43]

The Justice Department also refused to prosecute half a dozen oil companies involved in a scam to cheat the government out of $100 million a year in royalty payments through price-fixing and other scams. Under White House pressure, the Justice Department dropped most Clinton-era investigations of coal-fired power plants that were violating air quality standards. And in January 2003, the EPA and the Army Corps of Engineers sent out a memo forbidding enforcement personnel from ticketing polluters who fill or foul isolated wetlands without first clearing each case with Washington, D.C., headquarters—a policy designed to impede prosecution of law breakers.

Two days after Christmas 2001, with President Bush at his Texas ranch and most of official Washington on vacation, the White House announced that it was killing regulations that barred companies that repeatedly violate environmental and workplace standards from receiving government contracts. Up to $138 billion in federal contracts are awarded each year to companies violating environmental and labor safety statutes.

Ironically, despite its reluctance to enforce laws against polluters, the administration is still enthusiastic—even ingenious—about enforcing the law against environmentalists. In July 2003, John Ashcroft's Justice Department sought and obtained an indictment against Greenpeace for violating an obscure 1872 "sailor mongering" law designed to discourage flophouse owners from boarding ships to recruit sailors to drink at their taverns. The law had not been enforced in 132 years. The organization faced substantial fines, five years of federal probation, a criminal record, and enhanced punishment for future activities. In May 2004, a federal judge dismissed the case without even hearing the defense, recognizing it as a political prosecution aimed at injuring—and possibly shutting down—an organization that has been a thorn in the side of the administration's corporate paymasters.

Consistent with its twisted enforcement priorities, the Justice Department has awarded Wise Use founder Ron Arnold part of a $325,000 grant to study and report on environmental terrorism,[44] which Ashcroft has designated the country's top domestic terrorist threat,[45] ignoring the anti-abortion and right-wing terrorists who have killed hundreds of Americans.[46] No single action on the part of the Bush administration, however, has done more to advance the Wise Use agenda than the appointment of Gale Norton as secretary of the Interior. The Department of the Interior manages the richest treasure trove of all—450 million acres of public lands and 3 billion acres of coastal waterways. Turning Interior over to Norton was like handing the keys of the kingdom to the Wise Use camp. Norton, a champion of corporate welfare for three decades, is the very embodiment of Wise Use. One of the movement's most effective and fanatical leaders, as radical as James Watt, she is far more dangerous because she is attractive, charming, and diabolically clever.

Norton has gone to bat for polluters her entire career. In 1979, the year after she graduated from law school, Norton went to work for the Mountain States Legal Foundation as an attorney under the tutelage of then-director James Watt. Norton helped file lawsuits to dispute federal grazing limits, to impede EPA clean air rules,[47] and to support oil and gas drilling offshore, in wilderness areas, and in wildlife refuges.[48] Each of these lawsuits promoted the interests of the MSLF's major funders, including Exxon, Burlington Northern, the Independent Petroleum Association, and the Rocky Mountain Oil & Gas Association.[49]

In 1984, after a two-year stint at the Hoover Institute, Stanford University's right-wing think tank, Norton used her connections to gain positions first at the Department of Agriculture, then in the Interior Department, and finally on Reagan's Council on Environmental Quality. While working for the Pacific Research Institute for Public Policy in 1989, she acknowledged that her advocacy of "takings" legislation would mean the end of environmental enforcement. "I view that as something positive," she said. She served as Colorado attorney general from 1991 to 1999 and ran unsuccessfully for the Senate in 1996.[50]

She has served as an adviser to The Advancement of Sound Science Coalition (TASSC), the junk-science think tank led by Monsanto lobbyist Steven Milloy. TASSC receives funding from Philip Morris (to bless the safety of secondhand smoke) and from Exxon (to nay-say global warming), as well as from Procter & Gamble, Dow, and 3M.[51] She was a board member of the Wise Use group Defenders of Property Rights (DPR) and a Fellow at the Property and Environment Research Center (PERC), both recipients of funding from the right-wing Gang of Five. Among other things, PERC promotes views that are critical of recycling and skeptical of acid rain.

In 1998, Norton gained national prominence by founding her own organization, the Council of Republicans for Environmental Advocacy. She stacked its advisory board with powerhouse corporate crusaders like Newt Gingrich and got funding from Coors, Amoco, ARCO, the American Forest and Paper Association, and the Chemical Manufacturers Association.[52]

When she lets her hair down in front of Wise Use constituents, it becomes immediately clear that Norton is a genuine radical who despises not just government but the very idea of community. In a 1996 speech Norton bitterly shellacked federal incursions into what she considered the private domain. She listed the atrocities: the wheelchair ramp in the state capitol mandated by the Americans with Disabilities Act, the EPA's automobile emissions inspection program, federal requirements forcing schools to remove asbestos, the Fair Labor Standards Act, and the Violence Against Women Act.[53]

When Norton was appointed chief steward of America's natural resources, the Bush administration quickly placed a rapacious crew of Wise Use pirates in the key posts in her department. Deputy secretary and former mining industry lobbyist J. Steven Griles was joined by James E. Cason, whose position of associate deputy secretary was created to spare him the embarrassment of a Senate confirmation process. As an assistant secretary under James Watt, Cason had been condemned for giving millions of taxpayer dollars to the mining industry in a sweetheart deal. He had authorized a rule making national parks and wildlife areas vulnerable to strip-mining, and had claimed that the spotted owl would go on the endangered species list "over my dead body."[54]

The department solicitor was William G. Myers, executive director of a Wise Use group, Public Lands Council, an arm of the National Cattlemen's Beef Association that advocates grazing rights on government land for minimal fees.[55] H. Craig Manson became assistant secretary of the Interior for Fish, Wildlife, and Parks. When he was chief counsel to California governor Pete Wilson's Fish and Game Department, Manson sought to allow state officials to suspend California's Endangered Species Act under certain circumstances. His efforts were overturned by the California Appellate Court.[56]

Mining lobbyist Rebecca W. Watson became assistant secretary of Land and Minerals Management.[57] Mining lawyer and Wise Use leader Bennett Raley was named assistant secretary of Water and Science. Raley, who is in charge of ensuring the conservation of the nation's water supply, lobbied against the 1994 Clean Water Act reauthorization.[58]

Other appointees have shown similar contempt for the environment. Lynn Scarlett, now Interior's assistant secretary for Policy, Management, and Budget, was formerly president of the Reason Foundation, an industry-funded libertarian think tank that downplays the risks of global warming and air pollution. Scarlett opposes even the most innocuous initiatives, such as curbside recycling and nutrition labels on food.[59]

Jeffrey D. Jarrett, director of the Office of Surface Mining Reclamation and Enforcement, is an ex–coal company executive. In the early 1990s, as a midlevel manager for the office that he now runs, he was the subject of a criminal probe for obstructing justice.[60]

With this lineup in place, Norton opened up our public treasures to industry plunder. The list of environmental outrages already committed by Norton and her team is lengthy. She approved construction of the nation's largest pit mine in the foothills of Arizona's Gila Mountains, even issuing a statement claiming that the 3,360-acre copper mine would have no ecological impact. She has pushed several other catastrophic mine projects, including one on federal land a few miles from downtown Reno, another beneath the Cabinet Mountain Wilderness of northwest Montana, and a 1,571-acre strip mine on public land in California that is considered sacred by Native Americans. She canceled a ban on new mining claims on roughly 1.2 million acres in and around southwestern Oregon's Siskiyou National Forest. She overturned a Clinton-era regulation limiting the amount of public land that could be used for waste disposal of hardrock mining debris and announced new regulations that reverse environmental restrictions on mining for gold, copper, and other metals on federal lands.

Norton signed off on a plan to open nearly 9 million acres of Alaska's North Slope to oil and gas development, put the kibosh on a citizens' panel to oversee the trans-Alaska oil pipeline, brushed aside a record 25,000 opposing comments to approve a Houston company's request to embark on the largest oil and gas exploration project in Utah's history, and pushed for oil and gas drilling in Wyoming's Jack Morrow Hills.

She seems intent on turning some of the most beautiful landscapes in the country into oil fields, including parts of Padre Island National Seashore in Texas, the Canyons of the Ancients National Monument in Colorado, the Dome Plateau near Arches National Park in southern Utah's Redrock Canyon Country, the Upper Missouri River Breaks National Monument in Montana, and Big Cypress National Preserve in Florida. She also proposed redrawing the boundaries of America's national monuments to allow energy development.

Her record protecting wildlife is equally dismal. Hers is the first administration since passage of the Endangered Species Act to not voluntarily list a single species as threatened or endangered. In March 2003, just when gray wolves were beginning to recover out West, Norton's U.S. Fish and Wildlife Service proposed stripping federal protection to make them easier to kill. She shelved the 10-year Federal Salmon Recovery Plan adopted in 2000 and denied endangered status to imperiled species such as the California spotted owl and the cutthroat trout in Washington and the lower Columbia Basin along its border with Oregon. She moved to strip protection for the Imperial Sand Dunes Recreational Area near San Diego, as well as hundreds of thousands of acres in the Southwest that are home to the arroyo toad, the fairy shrimp, the endangered Quino checkerspot butterfly, and dozens of rare desert plants. Norton recommended that the Justice Department not appeal an Idaho court's ruling that denies water allocations to the Deer Flat National Wildlife Refuge on the Snake River in Idaho, depriving the river of water to support its booming trout fishery.

In April 2003, Gale Norton imperiled millions of acres of wilderness by signing a sweetheart deal with then-Governor Mike Leavitt of Utah that will make it easier for state and local officials to claim ownership of thousands of miles of dirt roads, trails, and fence lines under an obscure provision of federal law. That year, she also petitioned the United Nations to remove Yellowstone from a list of endangered World Heritage sites. And the list goes on.

Norton brooks no dissent. The Department of Interior, like every other federal agency, comprises long-term employees supervised by the current administration's appointees. The former provide continuity and support to each new administration and are often experts in the issues at hand. Norton, however, has transferred, fired, or demoralized many of them, and she has suppressed the findings of scientists in her own department. Her conduct has precipitated an epidemic of resignations, retirements, and whistle-blower lawsuits by high-level scientists, civil servants, and technical staffers.

She demands total obedience and resorts to brutal tactics when crossed. In December 2003, for example, National Parks Police Chief Theresa Chambers said, in a rather innocuous interview with the *Washington Post,* that increased security requirements are causing the park police to cut back on patrols.

Chambers' job demands that she speak to the press, the public, and Congress, but the Interior Department hierarchy wanted to require that she speak only with prior approval. After refusing to agree to a permanent gag order, Chambers was stripped of her badge and firearm with the recommendation that she be fired. As of this writing, she is still in legal limbo.[61]

I've had many brushes with Norton's crew of hardheaded ideologues. They are convinced that our government and its laws are illegitimate and that the illegitimacy makes it permissible for them to violate all the rules. I have seen them subvert the law, corrupt our democracy, and distort science. I have witnessed their willingness to break promises and deceive those they are appointed to serve.

In the summer of 2003, my cousin Maria Shriver's husband, Arnold Schwarzenegger, approached me out on Cape Cod. He was determined, he said, to be "the best environmental governor in California history." I agreed to help him and worked with a group of sympathetic Republicans and Democrats in California to draft Arnold's environmental platform. Among the key provisions was support for the Sierra Nevada Framework. The plan was the product of a decade of grueling work by government, the timber industry, and environmental groups to manage the Sierra Nevada forests. But Wise Use radicals at Interior opposed *any* restrictions on the exploitation of public lands.

Immediately after the election, David Drier, a conservative Republican congressman from California, asked Schwarzenegger, at the behest of the White House, to abandon support for the Framework. Schwarzenegger refused, but noted that if changes to the Framework were warranted by new information or science, the Framework should be modified by the same thoughtful, inclusive stakeholder process that had resulted in the original plan. This seemed to appease the White House. Karl Rove promised that no federal action would be taken on Framework protections, especially logging, without extensive discussions with the state and all stakeholders.

And yet, late in the afternoon of January 21, 2004, Governor Schwarzenegger received word that the U.S. Forest Service would announce a new plan for the Sierra Nevada, tripling logging levels over the Framework agreement. Schwarzenegger's office and California EPA commissioner Terry Tamminen tried frantically to reach administration officials, but all calls went unanswered. As if to emphasize its contempt for the process, the Forest Service held a press conference in the Sacramento Hyatt, directly across the street from the governor's office, to announce its plan. Republicans and Democrats alike in the Schwarzenegger administration were furious at the betrayal and astounded by such hardheaded arrogance. So much for states' rights.

There are several ways to measure the effectiveness of a democracy. One is to look at how much the public is included in community decision making. Another is to evaluate access to justice. The most telling aspect of a government, however, is how it distributes the goods of the land. Does it safeguard the commonwealth—the public trust assets—on behalf of the public? Or does it allow the shared wealth of our communities to be stolen from the public by corporate power? The environmental laws passed after Earth Day 1970 were designed to protect the commons. Since then, life has dramatically improved in America. Children have measurably less lead in their blood and higher IQs as a result. We breathe cleaner air in our cities and parks and swim in cleaner water in our lakes and rivers. These laws have protected the stratospheric ozone layer, reduced acid rain, saved threatened wildlife such as the bald eagle, and preserved some of the last remaining wild places that make this country so beautiful. In other words, they protect the America that we all hold in common.

But George W. Bush's policy advisers somehow don't see the benefits we've received from our investments in our country's environmental infrastructure. All they see is the cost of compliance for their campaign contributors—a group that is led by the nation's most egregious polluters. This myopic vision has led the White House to abandon its responsibility to protect the public trust.

Former Interior Secretary James Watt once promised, "We will mine more, drill more, cut more timber."[62] In April 2001, a retired James Watt told the *Denver Post,* "Everything Cheney's saying, everything the President's saying, they're saying exactly what we were saying twenty years ago, precisely. Twenty years later, it sounds like they've just dusted off the old work."

3

The First Round

During his presidential campaign, Bush threw a bone to environmentally conscious soccer moms and centrist Republicans. Global warming, he said in his second debate with Al Gore, "needs to be taken very seriously."[1] While he opposed the Kyoto Protocol,[2] the international agreement to slow down global warming, he proclaimed that under his leadership the United States would tackle the problem by strictly regulating CO_2, the principal greenhouse gas, projecting the image of a man determined to take a thoughtful approach to the environment.[3] As it turns out, he left this campaign promise on the stump.

Barely three months into office, Bush walked away from his pledge. It was, quite possibly, an unprecedented turnaround—I can't remember a president who's violated a major campaign promise so soon after his election, scorning the mandate that put him in office.[4] The move revealed the depth of industry clout at the White House. But as Bush and his advisers would learn, backpedaling on the environment doesn't play well in Peoria.

The "greenhouse effect" was first predicted in 1896, in a paper written by a Nobel Prize–winning Swedish chemist named Svante Arrhenius.[5] In 1988, NASA scientist James Hansen riveted the world when he testified before Congress that greenhouse gases were warming our climate with dire consequences for the future.[6] Since then we have developed better computer modeling and collected reams of scientific evidence—and seen 10 of the warmest years on record.[7] The ranks of the skeptics have thinned to a small army of industry-funded charlatans whose voices are amplified through the bullhorn of Rush Limbaugh and the shills at the Heritage Foundation.

Scientists agree that we are now pumping out vastly more CO_2 than the Earth's system can safely assimilate. The surplus gases create an invisible blanket in the atmosphere that prevents heat from being released to outer space, and as that heat builds up it changes the energy balance in the world, and that changes everything. During the Senate floor debate on the McCain-Lieberman Climate Stewardship Act in October 2003, Senator John McCain held up satellite photographs of the North Pole showing a dramatic 20 percent shrinkage of the Arctic Sea ice over the previous twenty-five years.[8] But you don't need a satellite to know the world is changing.

Glaciers are shrinking worldwide, except at the very highest altitudes. The mountain ranges from the Alps to the Rockies, the Himalayas to the Andes, are losing their snow pack, a trend already seriously impairing regional water supplies. The fringes of Antarctica's ice are melting. The 20,000-year-old permafrost of the northern tundra is softening. Ironically, Alaska's North Slope oil industry is being impeded by shortened operating seasons.[9] During the summer of 2003, 19,000 people died in Europe due to the hottest temperatures in at least 500 years.[10]

Temperatures are higher everywhere: in both hemispheres, on the earth's surface, within the soils, in the depths of the ocean, in the upper atmosphere, and within the ice sheets. Despite fluctuations, intense winter cold is less common in places like New York. Higher temperatures have increased evaporation from the ocean surface and intensified precipitation; gully-washer rainstorms, floods, and heavy blizzards are more frequent and destructive.[11] In North Carolina, the entire hog industry was built on the assumption that hundred-year floods came only once each century. But in the last ten years, three hundred-year floods have caused enormous economic and environmental woes. Warmer water is beginning to eradicate coral reefs worldwide—most of them may be gone by 2050.[12] Sea levels are rising, and coastal erosion is a growing crisis.

Even the timing of the seasons has begun to change. Ecosystems are starting to shift. Plants, animals, and insects are appearing in places they didn't before.[13] I regularly see black vultures now in upstate New York, but all my bird books describe its northern range as Virginia. Birds are laying their eggs at different times.

There is growing evidence that dramatic climate change may occur suddenly, a development that has even gotten the Pentagon's attention.[14] A report commissioned by Andrew Marshall, the father of Star Wars and the military's graybeard expert on future strategic threats, describes the human disasters that would occur if the climate shifted abruptly in a decade or two, as happened some 12,000 years ago. According to this scenario, most of Holland and Bangladesh would be submerged by violent storms and rising seas. Northern Europe would freeze because of disruptions to the Gulf Stream. Millions of environmental refugees would gather at the frontiers of the developed world, driven by wars, famines, and floods. Nuclear conflict, megadroughts, and widespread rioting would erupt across the world. Nations might be forced to expand their military power to defend dwindling food, water, and energy supplies. The report paints a picture of the United States as a giant gated community insulating itself from the world it helped create, isolated and despised by its angry, jealous neighbors. Although this outcome is presented as a worst-case scenario, climate change "should be elevated beyond a scientific debate to a U.S. national security concern."[15]

The first major international attempt to tackle global warming was the Rio Climate Treaty, signed by the first President Bush in 1992 and ratified by the U.S. Senate that year. The Rio treaty contained plans to return CO_2 and other greenhouse gas emissions to their 1990 level by 2000. In 1997, the Rio participants proposed a more detailed set of actions, the Kyoto Protocol, which requires that developed nations reduce their emissions of greenhouse gases 5 percent below the 1990 level during the period 2008–2012.[16]

President Bush's campaign promise to regulate CO_2 would have been a big step toward meeting the pledges that the United States made in the Rio treaty. Thanks to industry lobbying, CO_2 was not listed as a pollutant in the original Clean Air Act, so by 2000 it was still unregulated. The Clean Air Act, however, authorized the federal government to regulate *all* air pollutants, even those not specifically listed in the original act itself. The EPA had used this authority to develop regulations for toxic pollutants like lead, sulfur dioxides, and particulates that cause respiratory disease. Bush's promise indicated that he would add CO_2 to that list.

After Bush took office, his newly minted EPA director, Christine Todd Whitman, put global warming at the top of her agenda. The White House had sold the former New Jersey governor to the American public as an environmental moderate, citing her participation in a lawsuit by several eastern states against an Ohio Valley power plant.

From my front-row seat across the Hudson River, she had not seemed very moderate. She had signed on to that power plant lawsuit reluctantly. In fact, as governor from 1993 to 2001, she was one of the nation's leading advocates of pollution-based prosperity, cutting the state's Department of Environmental Protection budget by 30 percent, firing virtually all of the agency's enforcement attorneys, and relaxing enforcement of pollution laws in favor of "voluntary compliance" (which works about as well as "voluntary taxation"). She abolished New Jersey's renowned environmental prosecutor's office and dismantled some of the state's most important environmental laws. She replaced the state's public advocate with a business ombudsman and declared New Jersey "open for business." To the public and Congress, however, Whitman looked moderate next to the Wise Use fanatics that were being handed the other top jobs. Her nomination passed the Senate 99–0.[17]

Still, Whitman was not part of the Texas clubhouse crowd that formed the president's inner circle. As Ron Suskind reveals in his best-seller, *The Price of Loyalty,* only a favored few in the new administration had the ear of the president. A month into her appointment, Whitman was still trying to clarify the president's views on U.S. environmental policy.[18]

As a result, Whitman was unsure what message she should bring to her first major international meeting, in Trieste, Italy, in early March 2001. The meeting was intended to prepare the eight leading industrialized nations for the official Kyoto Protocol meetings that summer. Despite Whitman's abysmal environmental record, she recognized climate change as a genuine global crisis and, according to Suskind, she chose this issue to cultivate a reputation as an environmental steward.[19]

Whitman turned to Paul O'Neill at Treasury. Although energy and the environment aren't Treasury's principal bailiwicks, O'Neill could claim a credible expertise in global warming. He told me that in his former job as CEO of Alcoa, he had worked successfully to eliminate two potent greenhouse gases from the aluminum smelting process. "I once told my board, 'We are environmentalists first and industrialists second,'" he said. "I believed this in my heart and I had a place to act on my beliefs."[20]

In 1998, O'Neill had given a major speech on global warming to the aluminum industry that was later published as a booklet. Even then, he was frightened by the potential impacts of climate change. His overall message was that global warming, along with nuclear holocaust, must be taken more seriously than any other political issue. "If you get welfare reform wrong," he said, "you get a second chance—civilization doesn't go down the drain."[21]

Try as they might, however, neither O'Neill nor Whitman could engage the president on the topic. While O'Neill had weekly meetings with the president—more access than any cabinet official save Donald Rumsfeld—he came to feel that the president didn't listen or deliberate in an analytical way.[22]

According to Suskind, Whitman finally managed a meeting with Andrew Card and Condoleezza Rice to prepare a strategy on global warming. They agreed with her suggestions on what to say in Trieste: The White House was preparing to regulate carbon dioxide as a pollutant. That would make it appear that the United States, which emits 25 percent of the world's CO_2, was taking global warming seriously without committing the administration to the Kyoto Protocol. Whitman, confident that she had the White House's blessing, stated this position on CNN's *Crossfire* and in press conferences.[23]

Just days later, four right-wing Republicans with strong industry ties—Chuck Hagel of Nebraska, Wise Use icon Larry Craig of Idaho, South Carolina's Jesse Helms, and Pat Roberts of Kansas—sent a letter to the president, complaining about Whitman's press appearances and demanding a "clarification of your administration's policy on climate change." O'Neill told Suskind that he was almost certain the senators' letter had been prompted and possibly even written by Dick Cheney.[24]

Whitman, furious that she was being outmaneuvered, scrambled to schedule a meeting with the president. But the moment she arrived in the Oval Office, before she could even launch into her speech, Bush cut her off. "Christie, I've already made my decision." He held up a letter—a letter both O'Neill and Whitman believe that Cheney wrote—all ready to send back to the Republican senators, and began reading from it.

The president would oppose Kyoto, the letter said, because it exempted 80 percent of the world, including China and India, and it was an "unfair and ineffective means of addressing global climate change concerns." As for his campaign promise to regulate CO_2 in the United States, the letter explained that he had changed his mind. He argued that the Kyoto agreement would "cause serious harm to the U.S. economy," and emphasized the importance of energy development.[25]

As soon as Bush had read the letter, he signaled for her to leave. In a few moments, it would be sent to the senators and released to the world. Whitman was stunned. She wandered into the anteroom outside the Oval Office, Suskind reports, where one of the secretaries was handing a document to Vice President Cheney. "Mr. Vice President," the secretary said, "here's the letter for Senator Hagel." Dick Cheney picked up the letter, now freshly signed by the president, and brushed past Whitman on his way to Capitol Hill.[26]

As O'Neill related to Suskind, "It was a clean kill. She was running around the world, using her own hard-won, bipartisan credibility to add color and depth to his campaign pronouncements, and now she ended up looking like a fool."[27]

By March 2001, the president had officially walked away from the Kyoto Protocol, and the United States had pulled out of all debate and negotiations with the rest of the world on global warming. O'Neill and Whitman were not the only ones in shock. Some of my colleagues at the NRDC had worked for more than a decade with Republican and Democratic administrations, international scientists and environmentalists, and world leaders from over 100 nations to iron out an agreement that would balance environmental and economic needs against the planet's greatest crisis. Now the administration had not just changed the game plan—it had walked off the field. Its only proposal for dealing with global warming was denial. And its sole rationale appeared to be a desire to placate the coal and power industries and the Wise Use antiregulatory fanatics.

After the president's letter, the press wrote Whitman's political obituary, proclaiming her irrelevant. Some articles called for her resignation. The NRDC's political expert Greg Wetstone commented that Whitman "suffered the most immediate and visible loss of clout ever for a cabinet officer."[28] Even her fellow cabinet members openly joked about her diminished stature. "That's what Colin Powell had been calling me at cabinet meetings, the wind dummy," Whitman explained to a journalist. "It's a military term for when you are over the landing zone and you don't know what the winds are. You push the dummy out the door and see what happens to it."[29]

Whitman was demoralized but she wasn't about to step down. The Kyoto incident taught her a tough lesson: Industry was calling the shots, and if she didn't want to look like a feeble scold at a frat house orgy, she needed to toe the industry line. Within a week of the CO_2 fiasco, she found an opportunity to genuflect to industry and its Wise Use allies.

On March 20, 2001, Whitman quietly announced that the EPA would suspend a Clinton-era rule reducing the allowable amount of arsenic in public water supplies. Bush officials complained that the arsenic rule was a draconian standard that was hastily devised at the last minute by Clinton's people. While it's true that the arsenic rule wasn't finalized until the early days of January 2001, the rule was the result of years of wrangling between scientists, health experts, industry, and the public.

Several studies, including six by the National Academy of Sciences (NAS), indicate that arsenic is a potent carcinogen. Bush's problem, of course, is that much of the arsenic in our drinking water is the result of mining activities. And in 2000 the mining industry had shoveled $5.6 million into Republican Party campaign chests, with Bush receiving the lion's share of it.[30] He had also narrowly lost New Mexico, a big mining state, and the White House wanted to curry favor with the powerful mining interests there.

Whitman, aware that arsenic isn't exactly a crowd pleaser, faxed her press release to the New Mexico media and almost no one else. "They were hoping to avoid publicity," says Erik Olson, a lead attorney for the NRDC's public health program. The ploy backfired. Olson learned of the decision and immediately called the *New York Times.* "It's a huge deal," Olson told the *Times.* "This decision will force millions of Americans to continue to drink arsenic-laced water. A lot of people are going to die from arsenic-related cancers and other diseases if they weaken or delay this thing." The story made the front page of the national edition and kicked off a flurry of press and public outrage.[31]

Everyone from Leno and Letterman to a legion of cartoonists had fodder for the joke of the week. A cartoon by Mike Luckovich in the *Atlanta Journal-Constitution* had Bush telling Cheney, "I want arsenic in the water classified as a vegetable." *Doonesbury* had an empty cowboy hat explaining why the White House suspended the arsenic rules. "Fuzzy science, that's why. We need good strong science, good science is where our wings take dreams." Another panel had the Stetson explaining, "So, until we've really studied the polluted drinking water, I favor a voluntary approach." Question: "To cleaning it up?" Stetson: "No, to drinking it."

Meanwhile, 57,000 citizens and public interest organizations flooded the EPA and the White House with comments opposing the rollback of the arsenic rule.[32]

Fearing voter backlash in 2002, congressional Republicans were trampling one another to distance themselves from the president. The Republican-led House voted 218 to 189 in late July to block Bush and Whitman's initiative.[33] "They were getting calls in their own districts, and nobody really wanted to drink arsenic," said Olson.[34]

Then came the biggest blow of all: Citing, among other things, Bush's environmental policies, Senator James Jeffords of Vermont defected from the Republican Party, tipping the balance of power in the Senate to the Democrats. In his subsequent book *The Right Man: The Surprise Presidency of George W.*

Bush, the president's speechwriter David Frum called the arsenic decision "the worst blooper of the first year."[35] Bush later publicly acknowledged that the rollback was a political mistake. Whitman called it "a dumb decision—politically, really dumb."[36]

But Whitman justified the arsenic rollback, saying she wanted to verify the science behind Clinton's standard. She promised she would follow the science and quickly announced that she would convene two economic reviews of the standard, and yet another scientific study by the NAS (the seventh!).[37] The economic studies verified that Clinton's rule was fiscally sound. The NAS study was released on the evening of September 10. It was apparently not at all what Whitman had hoped: The NAS found that the EPA had *underestimated* the cancer risks of arsenic by about tenfold.[38] Whitman, who had pledged that her EPA would follow the science, was now faced with the prospect of dropping the standards even lower than the Clinton administration had suggested.

The next morning, the NAS study was a top story in the *Washington Post.*[39] The *New York Times* ran a 1,000-word article.[40] Olson was on his way to a D.C. press conference, happily considering the irony that the NRDC might get a lower arsenic standard from Bush than from Clinton.

Just before he arrived, American Airlines Flight 77 crashed into the Pentagon. The press conference was canceled, Washington shut down that day, and a new era in American history began. On Halloween 2001, Whitman quietly closed the comment period on arsenic before it was completed and announced that the EPA would not change the Clinton standard after all.[41] "The issue had become an albatross for Bush and 9/11 gave them a way to get out," Olson recalls. "I don't think anyone ever read the new NAS study. Everybody just forgot about it."[42]

The EPA gave industry five years to implement the ruling. The new standard will not take effect until 2006.[43]

Along with Bush's refusal to regulate CO_2, arsenic was a misstep that telegraphed where the administration was headed on the environment. Bush and his posse "came in with guns blazing, saying they wanted to repeal regulations, but they learned some hard political lessons quickly," says Reese Rushing, a policy analyst for OMB Watch.[44] They realized that the public cares deeply about environmental issues and that any open attempt to dismantle those protections would cause a backlash, particularly among women, a key voter group.

They wouldn't make the same mistake again. The anti-environmental agenda would not change, but the tactics would. The Bush White House had learned the value of stealth.

4

Cost-Benefit Paralysis

One of the most moving moments in my 20 years of environmental advocacy occurred at a December 2000 press conference in Raleigh, North Carolina. I had come to announce a Waterkeeper Alliance lawsuit against Smithfield Foods, whose industrial factory farms are polluting the waterways of North Carolina. I was standing at a podium in a large conference room overflowing with local press, when people I hadn't expected to see started filing in. They were black, white, men, women—tough and weather-beaten farmers, many in overalls, who had driven from all over the state to thank my Waterkeeper colleagues and me for standing up to an industry that had bullied them for years. Many came from families who had occupied the same piece of land for generations. Among them was 75-year-old Julian Savage from Bladen County, whose family had been in the area practically since the American Revolution. As I started to speak about the rural communities and once pristine waterways of eastern

North Carolina, and how they have been ruined by industrial pollution, I was astonished to see tears flowing down so many wrinkled faces.

One of the greatest sources of frustration to America's family farmers is public indifference to the cataclysmic struggle between traditional farmers and the industrial meat moguls. The farmers I faced that night had seen firsthand how meat barons bully vulnerable communities, trample their rights, and threaten their economic security. They understood better than anyone else that the consolidation of American food production by a tiny cabal of multinationals with no demonstrated loyalty to our nation or its laws threatens our food, our health, our culture—indeed, our very democracy.

The only hope these farmers had left was the American justice system. Over the years we have helped organize a national coalition of family farmers, fishermen, environmental and animal rights groups, religious and civic organizations, American Indians, and food safety advocates who are now fighting the hog industry in 34 states. I'm a specialist in this kind of litigation, which is why I ended up in Raleigh that day.

A year later, we finally had cause to celebrate. A Reagan-appointed federal judge ruled that the big factory farms were violating the Resource Conservation and Recovery Act (RCRA) and illegally operating without Clean Water Act permits. The legal decision sent a shock wave through the industry. But a few months after the ruling, the Bush administration intervened by issuing new regulations—written in concert with the industrial meat multinationals—that significantly weakened our lawsuit and ensured that hog factories would continue polluting rural communities across the United States indefinitely.

The demise of our case is a classic example of the White House's stealth strategy to dismantle America's environmental laws. Had the new rules run the gauntlet of public scrutiny, there would have been a howl of protest. Keeping the anti-environment maneuvers quiet was essential. And there was no better person to mastermind this strategy than John Graham, the director of the Office of Information and Regulatory Affairs (OIRA), an obscure agency inside the Office of Management and Budget.

Practically unknown outside the Beltway, OIRA's power is unmatched among federal agencies. Its official charter is to review every economically significant regulation proposed by the federal government and report the fiscal impacts to the White House. Federal departments and agencies develop these new regulations through an open process, guided by expert advice and mandatory public comment. Typically this takes six or seven years. Then, at the end of this highly democratic process, these regulations disappear into OIRA—only to emerge dramatically altered or not at all.

OIRA may be the most antidemocratic institution in government. It operates in secrecy. Congress created OIRA in 1980, but the agency reviews proposed regulations under an executive order, so most of its deliberations and records are inaccessible to the public.[1] Its decisions can profoundly affect the nation's health, safety, and environmental safeguards—unimpeded by public debate or accountability. OIRA's role depends largely on the White House. Under President Clinton, OIRA had a light touch. But in May 2001, President Bush nominated John Graham to head up the agency. The nomination horrified the environmental community. As the founder and director of an industry-funded think tank, the Harvard Center for Risk Analysis (HCRA), Graham has a long track record of ideological scorn for the public welfare. He is a favored guru of the Wise Use coalition and has been associated with several Wise Use think tanks and front groups, including the Mercatus Center, the Advancement of Sound Science Coalition, and the American Enterprise Institute.[2] He was also on the board of the American Council on Science and Health, which employs so-called tobacco science to defend a range of dangerous industry practices.[3] Among its media releases: "Why the National Toxicology Program Cancer List Does More Harm Than Good," "The Fuzzy Science Behind Clean-Air Rules," "Evidence Lacking That PCB Levels Harm Health," and, my personal favorite, "At Christmas Dinner, Let Us Be Thankful for Pesticides and Safe Food."[4]

Graham earned his stripes with his Wise Use allies by becoming the dark soothsayer in the occult art of cost-benefit analysis, a relatively new science that attempts to compare the costs of a proposed new law against its benefits to society. The first and most important environmental law, NEPA, requires that every government agency assess the costs and benefits of its actions and publish them in an environmental impact statement. But Graham takes this concept to extremes—reducing each cost and benefit to a number and then using a formula to determine public policy.

Even at its best, cost-benefit analysis has enormous potential for bias and distortion, allowing people like Graham to support almost any preordained conclusion. Graham's critics, including his academic colleagues, accuse him of routinely using discredited calculations and crooked methodologies to issue antiregulatory studies and pronouncements. "Figures don't lie," as the saying goes, "but liars figure." Graham, for example, often uses industries' own cost estimates, despite reams of studies showing that these numbers are grossly inflated.[5] His formulas almost always add up to the same thing: Industry wins, the public loses.

The usefulness of cost-benefit analysis is especially limited when it comes to environmental regulations. Cost-benefit analysis cannot possibly put monetary amounts on all the values of a healthy ecosystem. On the Hudson River I've seen the power industry use mathematical formulas to justify massive fish kills by measuring the value of the fish by the price of a fillet in a local grocery store. Since many species aren't fish that we typically eat, their destruction by the billions is valued at zero.

Such formulas also cannot calculate the way our community character is enriched by our environment. They cannot place a value on the Hudson River's 350-year-old commercial fishery, for example, which connects our children to their colonial and Algonquin heritages. Nor can they measure the aesthetic or spiritual dimensions of an unspoiled river. Patriotism, love of country, the sight of a bald eagle, the experience of a child catching fish, wading in the river and feeling the clean mud between her toes—all these things are connected to a wholesome environment. How do you put a value on human life, an unspoiled ecosystem, an unimpaired brain, and robust health without being subjective? OIRA's formulas mostly ignore the fact that human beings have other appetites besides money, and that if we don't fill them, we will never become the kind of beings our Maker intended. It's absurd to believe that one can reduce these things to numbers that are plugged into algorithms to dictate rational public policy.

Nor can cost-benefit analysis weigh the violation of basic human rights that are associated with environmental injury. The PCBs that we all now carry in our bodies, courtesy of Monsanto, General Electric, and Westinghouse, may or may not cause injury to a particular person. But each of us has the right to be free of that chemical trespass. My own PCB levels measure 1.83 parts per billion—well above average. No one has the right to put these chemicals in my body—or steal the air from my children's lungs—no matter how much they may profit by doing so.

John Graham's influence derives mainly from his exploitation of the Harvard name, which endows him with the gravitas and credibility that has eluded his fellow antiregulatory junk-science sorcerers. After publishing a risk-assessment study of automobile air bags in the early 1980s, he landed a job as an assistant professor in Harvard's School of Public Health, and soon big polluters found that they had a friend at Harvard.[6] "It turns out he was for sale," recalls Karl Kelsey, one of Graham's Harvard colleagues. Graham was so successful at using Harvard's name to push a corporate agenda that a deluge of industry money began flowing in. He launched the Harvard Center for Risk Analysis in 1989 and served as its director until 2001.[7] "It's horrifying," recalls Kelsey. "The man created this economic center that was funded up the wazoo, and you know, industry has been trying for years to buy out Harvard, and Graham was the first guy who would do it." Dean Harvey Fineberg pushed to grant Graham tenure and "the faculty had to go along," says Kelsey. When Graham was nominated to OIRA, Kelsey refused to sign a letter to the Senate by several of his colleagues opposing his confirmation. But after seeing the damage Graham has done to federal science since his appointment, Kelsey now regrets not having come forward earlier.[8]

The conflicts of interest at the HCRA are mind-boggling. From the start the center was funded by big polluters and trade associations representing the oil, chemical, auto, drug, agribusiness, and mining sectors with gripes against government regulations. They currently include Monsanto, Dow Chemical, E.I. DuPont, Exxon, General Electric, Union Carbide, Boise Cascade, the American Petroleum Institute, and the American Chemistry Council.[9] The HCRA's Advisory Council includes executives from DuPont as well as David Sigman, chief attorney for environmental affairs at Exxon Chemical Americas.[10] High-ranking officers from American Electric Power, the National Association of Manufacturers, Eastman Chemical, and Tenneco, Inc., sit on the HCRA's Executive Council.[11] Needless to say, consumer and environmental groups are not represented on the council.

As director of the HCRA, Graham played a central role in undermining support for many of the country's vital environmental protections through his "research" papers. His testimony before Congress and work with the media helped to endow anti-environmentalism with scientific gloss. The center's bimonthly journal, *Risk in Perspective,* for instance, routinely featured articles discounting the risks of children's exposure to pesticides or power plant emissions, almost always without disclosing the abundant funding it had collected from pesticide manufacturers and utility companies.

Graham's close working relationship with Philip Morris is a perfect illustration of the way in which he blatantly offered research assistance to companies that agreed to fund his organization. In the mid-1990s, while apparently collaborating with Philip Morris to undermine an EPA risk assessment of the cancer-causing effects of secondhand smoke, Graham accepted a $25,000 donation to the HCRA. He subsequently returned that check on the purported basis that HCRA was prohibited from accepting money from tobacco companies. In his letter enclosing the check, Graham suggested that the tobacconeer reissue the check from the account of one of its subsidiaries, Kraft Foods. Philip Morris happily complied. Graham's actions appear to have been aimed at hiding the fact that he received tobacco money. But the Kraft donation may have given Philip Morris more bang for its buck: Soon after receiving Kraft's money, Graham not only continued to work on the secondhand-smoke issue, but began soliciting Kraft's help for his work opposing restrictions on pesticide residues on foods.[12]

But even the Harvard name couldn't always protect Graham from his own fuzzy math. In March 1997, Graham argued in news appearances and before the National Transportation Safety Board that a new, unpublished HCRA report had convinced him that passenger-side air bags were not cost-effective since they cost $399,000 for each year of life saved.[13] After harsh criticism from auto safety advocates, Graham's study was peer-reviewed, resulting in a dramatic recantation by Graham in the *Journal of the American Medical Association;* it turned out that the true cost was $61,000 for each year of life saved.[14] Graham was forced to acknowledge that his own data revealed that air bags were a worthwhile investment. To this day, Graham refuses to disclose the amount of unrestricted funding from the auto industry to the HCRA.

Graham suffered a major embarrassment during his confirmation hearings for OIRA. He claimed in his résumé that a study on the cost-effectiveness of regulations was the primary basis for his reputation "as a scholar." The 1996 study, which included the examination of 90 environmental regulations, claimed that 60,000 lives could be saved if the government abandoned certain rules and reallocated resources more efficiently.[15] Graham had stated in appearances before Congress and elsewhere that his data showed that these federal regulations were guilty, in his own words, of the "statistical murder" of 60,000 Americans every year.[16] Thanks to Graham's aggressive promotion, his conclusion became a sacred tenet of industry, and his "statistical murder" statement was endlessly repeated by the radical right and routinely quoted by opponents of regulation. It is this study that established him as an expert and ushered him onto the national playing field.

The study, however, had never been peer-reviewed.[17] The peer review process, in which unbiased experts evaluate the author's data and verify the conclusions, is the gold standard in science; no reputable medical or scientific journal will publish a study that has not been peer-reviewed. Graham's OIRA nomination prompted Professor Lisa Heinzerling of Georgetown University's Law Center to take a closer look at his data. In testimony submitted to the Senate Governmental Affairs Committee, Heinzerling proved that Graham's work was completely misleading. Heinzerling revealed that in fact Graham had based his calculations on 79 environmental regulations that were never enacted. In his study, Graham presented the vast majority of environmental programs as if they had been implemented, yet only 11 of the 90 "inefficient" rules upon which Graham based his calculations had ever been enacted. Heinzerling also showed that in Graham's calculations, he grossly understated the value of a human life.[18]

Heinzerling's discovery of Graham's misleading conclusions forced him to recant the study, sheepishly admitting in testimony to Congress that his claimed 60,000 "murders" were an exaggeration.[19] In the end, his study, hailed as proof of regulatory inefficiency and cited dozens of times by the media and members of Congress as evidence of the need for regulatory reform, was worthless. Worse, the fact that Graham himself used his flawed study in a deceptive way in congressional testimony and other public appearances amounts to a devastating indictment of the way in which he regularly aids and abets his corporate paymasters.

Graham's screwball cost-benefit formulas, coupled with his well-documented industry bias, made his confirmation one of the more intense political battles of the spring of 2001. He was widely opposed by public interest groups, labor unions, and scientists. Two separate letters signed by dozens of academics—including 11 of Graham's colleagues from the Harvard Medical School and the Harvard School of Public Health, which houses Graham's center—were sent to the Governmental Affairs Committee to express opposition to the Graham nomination.[20]

The letters denounced Graham as a scientific charlatan and an ethical train wreck. "Professor Graham," the authors of one letter said, "has shown his willingness to over-ride health, safety, environmental, civil rights, and other social goals in applying crude cost-benefit tools far past the point at which they can be justified by existing scientific and economic data." The letter charged him with using "controversial risk management methodology" and "extreme and highly disputed economic assumptions" that "discount the real risks of well-documented pollutants such as dioxin and benzene." Graham, they warned, "has publicly rendered many opinions on complex and imperfectly understood scientific phenomena, such as the etiology of cancer and other diseases, despite his lack of a degree in the hard sciences." They accused him of routinely making deceptive comments "to the media and Congress" that undermine regulatory efforts "by understating many of the potential benefits of health, safety and environmental regulation and overstating their costs."

Observing that "Graham's work has, overall, demonstrated a remarkable congruency with the interests of regulated industries," these scientists condemned his ethical lapses: "We also have serious concerns about Professor Graham's disregard for widely accepted fundraising and research norms within academia. He has solicited and accepted unrestricted funds from corporations with a direct financial interest in particular regulatory issues addressed by his work, without acknowledging the role of his corporate benefactors."

They concluded that "Graham's record shows that he is unlikely to serve as an honest broker as OIRA director."[21]

Despite all that, Graham got the job. His narrow victory—he received 37 no votes in his Senate confirmation, the second highest of any Bush nominee after Attorney General John Ashcroft—was a triumph for the industry forces that had long supported his career.[22] With Graham in place, his Wise Use backers now had a man in Washington who could promote their agenda out of sight of public scrutiny.

Now, as the government's regulatory czar, Graham sits in judgment of virtually every new rule affecting workplace safety, consumer safety, public health, education, and the environment. Many proposed regulations disappear into the black hole at Graham's OIRA. In congressional testimony in March 2002, Graham boasted that since his appointment the previous July he had rejected more than 20 potential new rules, claiming inadequacies in cost-benefit analysis—more than were rejected in all eight years of the Clinton presidency. Not once has Graham's OIRA rejected a proposed rule for being insufficiently protective of public health, safety, or the environment. OIRA's clear overriding concern is cost to industry. A recent report by the United States General Accounting Office analyzed 85 federal agency rules that OIRA changed, returned, or withdrew between July 2001 and June 2002. The GAO found that OIRA had hobbled several proposed public health and environmental regulations, including more than two dozen rules suggested by the EPA.[23]

Graham seems hell-bent on demolishing as many existing regulations as he can. As soon as he was confirmed, he collaborated with corporate allies, including the American Chemistry Council, the Mercatus Center, and the American Petroleum Institute, to come up with a hit list of existing regulations to "reform." Graham presented 23 of those regulations to Congress the following December, recommending they be changed or eliminated. Among the rules Graham targeted were the EPA's plan to reduce arsenic levels in public drinking water, the preservation of roadless areas in forests, a proposed Interior Department rule prohibiting snowmobiles in National Parks, and rules controlling coal-burning power plants.[24]

One of those regulatory changes directly affected the Waterkeeper lawsuit that had brought me to that memorable press conference in North Carolina. The tensions behind that lawsuit had been simmering for decades. Industrial meat moguls and their agribusiness allies often argue that the disappearance of the traditional family farm is inevitable; the greater efficiencies of large operations, they claim, render the small business obsolete. Nothing could be further from the truth. The disappearance of the family farm has little to do with market forces. It is the direct consequence of government policies deliberately designed to favor agribusiness over traditional farmers. In no agricultural sector is this clearer than in meat production.

For 300 years, this country's family farmers produced more than enough beef, pork, and chicken for American consumers and export markets. They used traditional techniques of animal husbandry, recycling their manure to fertilize the soil to grow feed crops. They were proud stewards of their land and generally raised their animals in a humane manner. Study after study shows that these small operations are far more efficient than the giant farm factories.[25] But agribusiness has used its political and financial clout to eliminate agricultural markets, seize federal subsidies, and flout environmental laws to gain competitive advantage.

The fact is, an industrial meat factory cannot produce a pound of bacon or a pork chop cheaper than a family farmer without breaking the law. Several federal and state laws prohibit dumping waste into the environment. Traditional farms are exempt from these laws since they have enough land to use their manure as fertilizer to grow actual crops. Factory farms simply have too much waste per acre to put it to beneficial use. Hog barons build football-sized warehouses and cram genetically engineered hogs into tiny cages where they endure short, miserable lives deprived of sunlight, exercise, straw bedding, and interaction with other animals.[26] Concentrated waste from these facilities is saturated with dozens of toxic chemicals and antibiotics, which are fed to the pigs to stimulate growth and keep them from dying from the stress.

The waste stream from factory farms is prodigious, to say the least. A pig produces 10 times the fecal waste of a human being, and a facility with 50,000 hogs produces more waste than a city of half a million people.[27] Circle 4, a Smithfield facility licensed in Utah for 850,000 animals, can produce more waste every day than all the human beings in New York City combined. Of course, before it dumps waste into the water or onto the land, a city needs a permit that prescribes strict treatment requirements. And these facilities do, too. It says so in the Clean Water Act.

But costs of hauling or treating this excess waste in accordance with a permit would make it impossible for these corporations to compete with traditional family farmers. So the hog barons ignore the permit requirements. They simply dump the waste in open-air "lagoons" and spray it onto fields, saturating them with toxins in concentrations far greater than growing crops can assimilate. As a result, they transfer the cost of their production to the rest of us by polluting the land, air, and waterways.

Waste from industrial pork factories contains a witch's brew of nearly 400 toxic poisons, including heavy metals, antibiotics, hormones, deadly biocides, pesticides, and dozens of disease-causing viruses and microbes.[28] Millions of tons of this fecal marinade, produced by meat factories in 34 states, leaks or oozes into waterways or aquifers, or volatizes into the air. Hog factory contaminants have also fostered outbreaks of a previously unknown microbe, *Pfiesteria piscicida,* in U.S. coastal waters. *Pfiesteria,* "the cell from hell," kills millions upon millions of fish and causes pustulating lesions that won't heal, severe respiratory illness, and brain damage in humans who handle fish or swim in the water.[29] My clients include fishermen from the Neuse River in North Carolina, where an estimated 1 billion fish died from *Pfiesteria* in a single six-week period in 1991. Many of these men are covered with open sores and some can no longer manage their lives because of brain damage caused by *Pfiesteria.*[30]

The hog barons' business model relies on the assumption that they can evade prosecution for these crimes by improperly influencing government enforcement officials. Corporate meat producers intentionally locate their meat factories in poor and minority communities where they can crush and muzzle opponents. They routinely rely on intimidation and bullying lawyers to silence critics. They harass neighboring farmers for complaining to regulatory agencies about odors or water pollution or for participating in public hearings. The meat industry typically infiltrates state legislatures in order to manipulate local officials. In Iowa, North Carolina, Michigan, and other states, legislatures have stripped local officials of their decision-making powers so that hog factories cannot be zoned out by local planning boards or health officials. When staff members from the Leopold Center for Sustainable Agriculture, a state-financed organization that advocates for family farmers, attended my Waterkeeper Alliance seminar on factory farming in Iowa in the spring of 2002, the pork barons got the Iowa State Legislature to retaliate by cutting $1 million, or 85 percent of the center's annual budget.

Believe it or not, the industry's capacity to corrupt legislatures is so vast that it has been able to persuade legislators in Missouri and Illinois to pass laws making it a crime to photograph factory farm animals. Thirteen states now have veggie "libel" laws that make it illegal to criticize food from factory farms and other agribusiness products. Remember folks, this is America!

In 2001, the Pork Producers Council launched a smear campaign against the Waterkeeper Alliance and against me personally. The industry created a Wise Use front group, "Truthkeepers," whose sole purpose appears to be to discredit me. The president of Truthkeepers is an industry goon named Trent Loos, a former South Dakota pork factory manager. Loos follows me across the country, shadowing me at public appearances in New York, California, Illinois, and Kansas. Sporting a handlebar mustache, cowboy boots, a dandy scarf, and, appropriately, a black Stetson, he sits in the front row at my speeches and conferences, holding a microphone and a tape recorder. In Minnesota he followed me into a restroom and lurked behind me at the urinal. In Iowa he waited for me in a dark parking lot and made menacing comments. At an agricultural conference in Gettysburg, Pennsylvania, he trapped me in a hotel corridor, and I had to threaten fisticuffs to access my room. I was recently relieved of Trent's company when he was charged with cattle theft in Nebraska.[31] He pleaded guilty, and his probation prohibited him from leaving the state.

Waterkeeper Alliance was led into the hog-industry battle by Colonel Rick Dove, a commercial fisherman and 27-year Marine Corps veteran who had become a Riverkeeper on the Neuse after pollution from industrial hog factories put him out of business. Dove had been engaged in nose-to-nose combat with Smithfield, the Pork Producers Council, and the American Farm Bureau, industrial meat's most powerful ally. Masquerading as an ally of small farmers, the American Farm Bureau is a multibillion-dollar lobbying conglomerate of insurance, chemical, and agribusiness interests.

Following the December 2000 press conference, Waterkeeper sued four of Smithfield's North Carolina factory farms under the Clean Water Act and the Resource Conservation and Recovery Act, which forbids dumping pollutants onto soils. According to these statutes, anyone discharging waste into water or onto land needs a permit. Such permits require the dischargers to first treat their waste so that it will not injure public health or the environment. Smithfield countered with a motion to dismiss, arguing that it was entitled to the same exemptions that small independent family farms get.

But the Honorable Judge Malcolm Howard, a federal magistrate appointed by Ronald Reagan, sided with us. On September 20, 2001, Judge Howard issued a decision that sent a shock wave through the industry. Pig factories, the ruling implied, are no different than any other factories that dump waste. If they don't have a Clean Water Act permit, they are operating illegally, which meant that almost none of Smithfield's approximately 1,500 company-owned or -controlled facilities were in compliance with federal law.

A few weeks later, in November 2001, OIRA administrator John Graham and his staff held a meeting with lobbyists from industries affected by Judge Howard's decision: the Pork Producers Council, the National Turkey Federation, Farmland Industries, the National Cattlemen's Beef Association, the National Milk Producers Federation, and the United Egg Producers. Graham also invited the Mercatus Center to come to the rescue. Mercatus's eight-member board includes Frank Atkinson, a partner in McGuireWoods, LLP, the law firm defending Smithfield in our case before Judge Howard. Mercatus presented Graham with a draft of sweetheart regulations that would alter the Clean Air and Clean Water Acts and dramatically affect Judge Howard's decision. Graham sent the proposals to the EPA for review, ordering the agency to give them "high priority."[32]

The EPA was primed to do Graham's bidding. Among the EPA team was Adam Sharp, associate assistant administrator for Prevention, Pesticides and Toxic Substances. Sharp was a former official with the American Farm Bureau who has testified before Congress opposing air quality regulations for industrial agriculture."[33] The EPA developed a new set of regulations along the lines of Graham's recommendations.

Smithfield couldn't have asked for more. One key provision allowed the company to shirk responsibility for getting rid of most of its waste: Big "integrators" like Smithfield hire contractors to raise their animals. Smithfield, for example, owns approximately 260 factory farms in North Carolina but operates another 1,200 that are owned by desperate contract farmers under tight control by Smithfield. Under the contract terms, Smithfield owns the pigs and feed, while the unfortunate farmer owns only the mortgage on the hog house—built to Smithfield's specifications—and the manure. It's up to the farmer to get rid of the waste. When it ends up in the air, water, and groundwater, the state has no recourse, since the farmer has no money.

The agency also dropped the provision requiring meat factories to monitor groundwater, leaving communities vulnerable to nitrogen and other contaminants. Elevated nitrogen levels can cause blue baby syndrome or death in infants. In addition, the EPA removed a provision that required factory farms to consult with certified specialists in developing "nutrient management plans," and instead allowed polluters to write their own plans without oversight. It also altered the standards so that these plans, which are effectively a license to pollute, don't have to be made available for public notice and comment.

After signing off on Graham's recommendations, the EPA formally submitted the new regulations to OIRA in September 2002. Documents from the agency's rule-making record indicate that even those weakened rules were battered to death during OIRA's final review, which lasted three months. During this time, OIRA staff again met with a number of affected industry groups, including the American Farm Bureau, acquiescing to its request to leave the issue of dumping manure to the states.

The new regulations, in a coup de grace, decreed that tons of toxic waste oozing from giant lagoons were not subject to the Clean Water Act.

"The Clean Water Act was one of the few tools that small farmers had to take on the big multinational industrialists," David Friedman, the president of the American Farmers Union, the country's largest organization of independent family farmers, told me. "Now the Bush White House has stolen it from us."

As soon as the new rules were issued, Smithfield went out and got permits for all of its facilities. The permits are illegal under the Clean Water Act, and the NRDC and Waterkeeper are suing the EPA. But the meat bullies have won many years of reprieve, a result that went far beyond their wildest dreams.

As I drive through industrial farm country with Rick Dove, I can see the desperate, ruined communities the meat moguls have left in their wake: deserted main streets; bankrupted hardware and feed stores; banks, churches, and schools shut down. The heartland and its historic landscapes are being emptied of rural Americans and occupied by corporate criminals. These industries are murdering Thomas Jefferson's vision of an American democracy rooted in tens of thousands of independent freeholds owned by family farmers, each with a strong personal stake in our system.

In violating our federal antipollution laws like the Clean Air and Clean Water Acts, the meat barons are stealing something that belongs to the American public—the purity of our land and waterways, healthy air, abundant fisheries. They have lulled people into accepting unhealthy, unsavory meat and have gained a stranglehold on U.S. commodity production that poses a genuine threat to our food independence and national security.

But our battle is more than a fight to save American farmers and fishing communities. It is a fight to preserve human dignity. Like some deadly spider, John Graham sits in his secretive office and spins a dangerous web, weaving backroom deals with industry lobbyists and making a mockery of democratic government.

5

Science Fiction

As Jesuit schoolboys studying world history, we learned why Copernicus and Galileo kept their discoveries under wraps: a less restrained heliocentrist, Giordano Bruno, was burned alive in 1586 for the crime of sound science. With the encouragement of our professor, Father Joyce, my classmates and I marveled at the capacity of human leaders to corrupt noble institutions. Lust for power had caused the Catholic Church to ignore its most central purpose: the search for existential truths. It was my first exposure to the idea that money and political power can trump science.

Today, the Bush administration and Congress are similarly twisting science to consolidate power. John Graham's schemes to massage data and pervert the regulatory process in the inner sanctum of OIRA are part of a larger plan. To justify its agenda, the Bush White House is suppressing studies, purging scientists, and doctoring data to bamboozle the public and press. It is a campaign to suppress science arguably unmatched in the western world since the Inquisition.

In his infamous 2003 memo to his Republican brethren, GOP strategist Frank Luntz warned that the environment is the party's Achilles' heel.[1] He urged his pals to change their rhetoric. "Climate change," he wrote by way of example, "is less threatening than global warming. While global warming has catastrophic connotations attached to it, climate change suggests a more controllable and less emotional challenge." Luntz's advice to GOP leaders seems clear: Find scientists willing to hoodwink the people. "You need to continue to make the lack of scientific certainty a primary issue," he wrote, "by becoming even more active in recruiting experts sympathetic to your view." In that banal sentence, Luntz summed up the ethos of the administration he serves.

Science-bashing is nothing new. During my early days as an environmental litigator prosecuting polluters on behalf of Hudson River fishermen, I grew accustomed to seeing industry money corrupt talented scientists. But a succession of Republican-controlled Congresses has allowed this practice to infiltrate our government. When faced with a mounting body of evidence that runs counter to the interests of major campaign donors, sympathetic senators and congressmen parrot the industry complaint that "more study is needed," while wielding budget knives to see that more study never occurs. The Bush administration has followed suit in slashing funding for environmental science. The president's proposed budget calls for double-digit cuts in research at the EPA, the U.S. Geological Survey, the Department of Agriculture, and the Energy Department's Office of Science, among others.[2]

The anti-environmental zealots have at their disposal Wise Use's lavishly funded brigade of hired guns and "biostitutes"—crooked scientists on industry payroll, housed in fancy think tanks that publish junk science—to persuade the public that there are no environmental crises and to undo the laws challenging their pollution-based profits. They argue that pesticides are harmless; that global warming is a myth; that Mount Pinatubo, not chlorofluorocarbons, caused the ozone hole; that clear-cutting is good forest management; and that Alaska's caribou love the pipeline.

"At the shallowest level it's a cheap deception of the general public," says Princeton University geo-scientist Michael Oppenheimer. "At its worst, this approach represents a serious erosion in the way a democracy deals with science. You create high-sounding credentials and talk in tones that seem scientifically sensible, while all the time you are just fronting for a political agenda."[3]

I have had my own experiences with Torquemada's successors in the White House. At the time of the September 11 attacks in New York and Washington, I had just opened an office at 115 Broadway, catty-corner to the World Trade Center. When my partner Kevin Madonna returned to the office in November, he suffered a burning throat, nausea, and a headache that was still pounding 24 hours after he left the site. Despite the EPA's claims that the air was safe, Kevin refused to return, and we closed the office. Many workers did not have that option; their employers relied on the numerous EPA press releases between September 15 and December reassuring the public about downtown Manhattan's wholesome air quality. On September 18, none other than EPA administrator Christine Todd Whitman proclaimed, "I am glad to reassure the people of New York and Washington, D.C., that their air is safe to breathe."[4]

Not everyone bought the party line. New York's Senator Hillary Clinton and Congressman Jerrold Nadler, whose district encompasses the World Trade Center site, asked the EPA's ombudsman office to look into the matter.[5]

The ombudsman's office is an independent complaint department within the EPA whose function is to give the public a voice in cleanups of major hazardous waste sites that otherwise might devolve into backroom deals between regulators and polluters. Ombudsman investigators have a bloodhound's nose for corruption, and the stench at the World Trade Center site set them to howling.

In particular, they knew that Christine Todd Whitman was juggling some heavy-duty conflicts of interest: Whitman's husband has a deep and continuing financial involvement with Citigroup, which owns Travelers, one of the insurance companies responsible for compensating victims of the attack.[6] Citigroup stood to save hundreds of millions of dollars from Whitman's assurances about safety: The faster people went back to their homes, the less Travelers would have to pay for alternative housing. Whitman and her husband were also major bondholders in the New York–New Jersey Port Authority, which owns the World Trade Center and might benefit from downgrading the risks.[7]

The EPA's ombudsman at the time was Robert Martin. He appointed a 30-year solid-waste veteran, Hugh Kaufman, a master engineer and policy analyst for the EPA's Office of Solid Waste and Emergency Response, to work on Nadler and Clinton's complaint. Kaufman and Martin discovered that Whitman had downplayed the risks of Ground Zero to the point of lying. They interviewed a large group of EPA employees and other scientists who felt that Ground Zero was far more contaminated than many Superfund sites where respirators and moon suits are mandatory. They were alarmed that government officials were not advising appropriate precautions.[8] When Juan Gonzalez of the *Daily News* started reporting Martin's and Kaufman's findings, the EPA blasted the claims as "irresponsible."[9] In November 2001 Whitman removed Kaufman and Martin from the case and issued an order closing the ombudsman's office.[10] During a weekend in the following April, she sent five agents to confiscate Martin's files and padlock his office. Exhausted from the battle, Martin resigned, hoping that Congress would step in.[11] He's still waiting.

Three months later Kaufman won a ruling by the Department of Labor, which found that there was "no evidence of a valid reason for his removal" and ordered him reinstated.[12] Despite Kaufman's victory the independent ombudsman's office was effectively abolished.

The cat was out of the bag, however, and in August 2003, another watchdog within the EPA, the Office of the Inspector General, finally released a report that condemned the administration's handling of the aftermath of the World Trade Center attacks. The inspector general's report, based on the damning documents assembled by Kaufman and Martin, found that on the day that Whitman declared the air "safe," the EPA had not yet received the results of the first tests for toxins like cadmium, chromium, dioxin, or PCBs.[13] Days after the attack, the EPA announced that asbestos dust in the area was very low or entirely absent. In fact, more than 25 percent of the samples that the agency had collected around that time showed the presence of dangerous levels of asbestos.[14]

The IG report found that White House officials had altered language in the EPA's news releases to make them less alarming, pressuring the EPA "to add reassuring statements and delete cautionary ones." The White House blocked public access to raw data from the EPA's air testing and ordered the agency to delete warnings advising "sensitive populations" to avoid exposure and reword its directive that all residents in the area have their apartments professionally cleaned of toxic dusts. The White House forced the EPA to add language to a press release announcing that "our tests show that it is safe for New Yorkers to go back to work in New York's financial district" at a time when the EPA's tests were showing levels of asbestos 200 to 300 percent above those considered safe by the agency. The EPA associate administrator admitted to the inspector general that the desire to reopen Wall Street was a consideration when the press releases were being prepared. "EPA's basic overriding message was that the public did not need to be concerned about airborne contaminants caused by the World Trade Center collapse," says the IG report.[15]

A subsequent newspaper story described "screaming telephone calls" between EPA associate administrator Tina Kreisher and Sam Thernstrom, communications director for the White House Council on Environmental Quality. Kreisher, who now works as a speechwriter for Gale Norton, later acknowledged that she "felt extreme pressure" from Thernstrom.[16] Thernstrom's boss was Council on Environmental Quality director James Connaughton, a former asbestos-industry lawyer who had left industry for "public service" three weeks before. According to the inspector general's report, Connaughton did not want data on health hazards given to the public.

The government's reassurances may have endangered the health of firefighters, police, construction workers, and residents, including schoolchildren. Testing by outside sources has since revealed that contaminant concentrations at Ground Zero were among the highest ever recorded. For example, scientists from the University of California at Davis, who had conducted over 7,000 similar tests at contaminated sites worldwide, found particulates at levels they had never before seen.[17] One study done by Mount Sinai Medical Center in New York found that 78 percent of rescue workers suffered lung ailments and 88 percent had ear, nose, and throat problems in the months following the attack; half of those still had persistent lung and respiratory symptoms 10 months later.[18]

Dan Tishman, whose company, Tishman Construction, was involved in the reconstruction at 140 West Street, required his crews to wear respirators, but he recalls seeing many rescue and construction workers laboring unprotected—no doubt relying on the government's assurances. "The frustrating thing," Tishman lamented to me, "is that everyone just counts on the EPA to be the watchdog of public health. When that role is compromised, people can get hurt."[19]

"In the World Trade Center, the White House and the EPA were basically lying to the people of New York," Kaufman said. "It's public be damned at the EPA."[20]

Alas, this was not merely a desperate measure taken at a desperate time; this White House routinely goes to great lengths to withhold vital health information from the public. In May 2003, it blocked the EPA staff from publicly discussing contamination by the chemical perchlorate—an ingredient in solid rocket fuel. In an apparent effort to please defense contractors, the administration also froze federal regulations on perchlorate, even as new research revealed that alarmingly high levels of the chemical—which can compromise fetal development—had been detected in water in more than 20 states.[21]

For nine months the White House Office of Science and Technology sat on a report exposing the frightening impact of mercury on our children's health, finally releasing it in February 2003.[22] Among the report's findings was the disturbing fact that the bloodstreams of 1 in 12 American women are coursing with enough mercury to cause neurological damage, permanent IQ damage, and a grim inventory of other diseases in their unborn children.[23] (A more recent EPA study has found that 1 in 6 women carry dangerous levels of mercury and that some 630,000 children born each year are at risk.[24])

Then, in March 2004, the administration compounded that cover-up by helping the tuna industry conceal from consumers the true extent of mercury contamination in fish. Under pressure from tuna-industry lobbyists, the EPA and the FDA issued, instead, a mild warning about fish consumption by young children and women of childbearing age, after rejecting the recommendation of an FDA advisory committee. The advisory gently observes that albacore tuna "has more mercury than canned light tuna" but does not discuss frightening results of recent tests by the FDA that found canned albacore tuna to have about three times the mercury of canned light tuna, which itself is too contaminated for children or fertile women to eat frequently.[25]

One member of the advisory committee, University of Arizona toxicologist Vas Aposhian, quit in protest, pointing out that the panel of experts had advised warning children and childbearing women not to eat albacore tuna at all and to eat less light tuna than allowed by the advisory.[26] "What is more important to the U.S.," Aposhian asks, "the future mental health of young American children or the albacore tuna industry?"[27]

But of all the debates in the scientific arena, however, there is none in which the White House has cooked the books more than that of global warming. In the past three years the White

House has altered, suppressed, or attempted to discredit close to a dozen major reports on the subject. These include a 10-year peer-reviewed study by the Intergovernmental Panel on Climate Change, commissioned by the former president Bush in 1993 in his own effort to dodge what was already a virtual scientific consensus blaming industrial emissions for global warming. The list also includes major long-term studies by the federal government's National Academy of Sciences, the National Oceanic and Atmospheric Administration (NOAA), and the National Aeronautics and Space Administration (NASA), as well as a 2002 collaborative report by scientists at all three of those agencies.

In September 2002, administration censors released the annual EPA report on air pollution without the agency's usual update on global warming, that section having been deleted by Bush appointees at the White House.[28] On June 19, 2003, a State of the Environment report commissioned by the EPA in 2001 was released after language about global warming was excised by flat-earthers in the Bush administration. The deleted passages had included a 2001 report by the National Research Council, commissioned by the White House. In its place was a reference to a propaganda tract financed by the American Petroleum Institute.[29]

In July 2003, EPA scientists leaked an analysis that the agency's leadership had withheld for months showing that a Senate plan to reduce the pollution that causes global warming could achieve its goal at very small cost.[30] Bush reacted by launching a 10-year, $100 million effort to prove that global temperature changes have, in fact, occurred naturally, another delay tactic for the fossil fuel barons at taxpayer expense.[31]

"This administration likes to emphasize what we don't know while ignoring or minimizing what we do know, which is a prescription for paralysis on policy," says Princeton's Oppenheimer. "With a president who does not believe in evolution, it's hard to imagine what kind of scientific evidence would suffice to convince him to take firm action on global warming."[32]

When the administration can't actually suppress scientific information, it simply issues a new set of facts. The White House has taken special pains, for example, to shield Vice President Dick Cheney's old company, Halliburton, from sound science. The company is the leading practitioner of hydraulic fracturing, a new process used to extract oil and gas by injecting benzene into underground formations. EPA scientists studying hydraulic fracturing in 2002 found that it could contaminate groundwater supplies. A week after reporting their findings to congressional staff members, however, the EPA revised the data to indicate that benzene levels would not exceed government standards.[33] In a letter to Congressman Henry Waxman (D-California), EPA officials said the change was based on "industry feedback."[34] Waxman requested a clarification on the nature of that feedback. He's still waiting for an answer.

Interior Secretary Gale Norton seems particularly inclined to manipulate science. In autumn 2001, she testified that Arctic oil drilling would not harm hundreds of thousands of caribou. Not long afterward, Fish and Wildlife Service biologists contacted Public Employees for Environmental Responsibility (PEER), which defends scientists and other professionals working in state and federal environmental agencies. "The scientists provided us the science that they had submitted to Norton and the altered version that she had given to Congress a week later," says Jeff Ruch, the group's executive director.[35]

There were 17 major substantive changes, all of them downplaying the reported impacts to the caribou. When Norton was asked about the alterations in October 2001, she dismissed them as typographical errors.

During the late winter and spring of 2002, Norton and White House political adviser Karl Rove pressured National Marine Fisheries scientists to alter findings in a report on salmon in Oregon's Klamath River.[36] The final report underestimated the amount of water needed to keep the salmon population alive in order to divert a bigger share of the river water to large corporate farms. Agribusiness got its water and more than 33,000 chinook and coho salmon died—the largest recorded fish kill in the history of the American West.[37] Mike Kelly, the biologist who worked on the original draft—and who has since resigned in despair—told me that the coho salmon is probably headed for extinction. "Morale is low among scientists here," Kelly says. "We are under pressure to get the right results. This administration is putting the species at risk for political gain—and not just on the Klamath."[38]

Norton has also ordered the rewriting of an exhaustive 12-year study by federal biologists detailing the injuries that Arctic drilling would impose on populations of musk oxen and snow geese. She reissued the biologists' report as a two-page paper showing no negative impact to wildlife.[39] She ordered suppression of two studies by the Fish and Wildlife Service, concluding that the drilling would threaten polar bear populations and violate an international treaty protecting the bears.[40] She suppressed findings that mountaintop mining would cause "tremendous destruction of aquatic and terrestrial habitat." She forced Park Service scientists to alter a $2.5 million environmental impact statement that found that snowmobiles were damaging Yellowstone's air quality, its wildlife, and the health of its visitors and employees. A federal judge has reprimanded her office for interfering with scientists in the case.[41]

Roger G. Kennedy, former director of the National Park Service, views Norton's machinations with horror. "This administration routinely mismanages scientific information through distortion and omission whenever scientific truth is inconvenient to its industrial allies," he says. "It's hard to decide what is more demoralizing about the administration's politicization of the scientific process—its disdain for professional scientists working for our government or its willingness to deceive the American public."[42]

Manipulating data leads to one pesky problem: scientists who stick to their guns. And when scientists resist the White House agenda, the Bush camp threatens, intimidates, or purges them. Nearly every week I come across courageous public servants like Mike Kelly, or David Lewis, an EPA scientist for 30 years who endured reprisals for divulging that his agency knowingly relied on faulty data to approve the use of dioxin-tainted sewer sludge as farm fertilizer.

In April 2002, James Zahn, a nationally respected microbiologist with the Department of Agriculture's research service in Ames, Iowa, accepted my invitation to speak at a conference of over 1,000 family farm advocates, environmentalists, and civic leaders in Clear Lake, Iowa. In a rigorous, taxpayer-funded study, Dr. Zahn had identified bacteria that can make people sick—and that are resistant to antibiotics—in the air surrounding industrial hog farms. His studies proved that billions of these "superbugs" were traveling across property lines daily, endangering the health of neighbors and their livestock.[43] I was shocked when Dr. Zahn canceled his appearance on the day of the conference under orders from the U.S. Department of Agriculture in Washington. I later uncovered a fax trail proving that the order was prompted by lobbyists for the National Pork Producers Council. Dr. Zahn told me that his supervisor at the USDA, under pressure from the hog industry, had ordered him not to publish his study, and that he had been forced to cancel over a dozen public appearances before local planning boards and county health commissions seeking information about the health impacts of meat factories. Soon after my conference, Zahn resigned from the Department of Agriculture in disgust.[44]

On April 19, 2001, at the request of ExxonMobil, the Bush administration orchestrated the removal of Dr. Robert Watson, the NASA atmospheric chemist who headed the United Nations Intergovernmental Panel on Climate Change (IPCC).[45] Watson, one of the world's top climate experts and a key player in the global response to climate change, was despised by many in the oil and coal industries. He was replaced by a little-known scientist from New Delhi, India, who would be generally unavailable for congressional hearings.

When a team of government biologists indicated that the Army Corps of Engineers was violating the Endangered Species Act in managing the flow of the Missouri River, it was replaced by an industry-friendly panel. The original team had recommended that the Corps-controlled dam flows should be changed to mimic the natural pattern of rising in the spring and ebbing in the summer, a plan also endorsed by the National Academy of Sciences. Allyn Sapa, a former Fish and Wildlife Service supervisor in North Dakota and a member of the original team of scientists, says that there was "intense political pressure" brought to bear to maintain the status quo to please downriver agribusiness and barge operations. Sapa worries that the status quo may mean extinction for Missouri River species, including the pallid sturgeon. "We might be in the last decade for these fish," he says. "They can maybe make one or two more spawning runs."[46]

In April 2003, the EPA suddenly dismantled an advisory panel composed of utility industry representatives, state air-quality officials, scientists, and environmentalists who had spent nearly 21 months developing rules for stringent regulation of industrial emissions of mercury. John A. Paul, supervisor of Ohio's Regional Air Pollution Control Agency and the panel's cochair, says, "You have an EPA that assumes that because the law has an adverse impact on industry profits, the agency must find a way to usurp the law."[47]

The replacement of accomplished scientists and public health leaders with industry-friendly representatives has ignited grave concern among public health professionals. When, for example, the Centers for Disease Control and Prevention recently replaced an environmental health advisory panel with industry representatives (including a vice president of the Heritage Foundation), ten leading scientists denounced the move in the journal *Science.* "Scientific advisory committees do not exist to tell the secretary what he wants to hear but to help the secretary, and the nation, address complex issues," they asserted. "Regulatory paralysis appears to be the goal here, rather than the application of honest, balanced science."[48]

Needless to say, such appointments pose a direct threat to our children's health. In the summer of 2000, officials at the Centers for Disease Control asked Bruce Lanphear, along with four other esteemed health professionals, if he was willing to be nominated to serve on the CDC's Advisory Committee on Childhood Lead Poisoning Prevention. Lanphear, a physician, is director of the Environmental Health Center at Cincinnati Children's Hospital Medical Center and a professor at the University of Cincinnati. Approval of such nominations by the secretary of Health and Human Services is traditionally a formality. But before this could happen, George W. Bush took office.

A month later, the CDC turned down Lanphear's nomination, along with the others, selecting instead five people with ties to the lead industry. Among the new appointees are Dr. Kimberly Thompson, an associate of John Graham's Harvard Center for Risk Analysis. At the time of her appointment, the HCRA had 22 corporate backers with a financial interest in looser lead standards. Another choice was Bill Banner, who works part-time as an expert industry witness for lead poisoning cases, where he testifies that there is no harm to children from lead blood levels below 70 micrograms per deciliter.[49] This is dangerous nonsense. "We know kids often die if their blood levels get much above 80," Lanphear says. "At levels of about 10, which is about 3 percent of kids, we're pretty convinced they're harmful. Now there are several studies coming out saying there should be no threshold, that there are adverse effects below 10."[50]

"The current Bush administration has taken intolerance of science dissent to a new orbit," says Tom Devine, a self-described Goldwater Republican who monitors the treatment of federal whistle-blowers for the nonpartisan Government Accountability Project. "The repression against internationally renowned professionals and experts in their fields just for exercising the scientific method objectively is unprecedented."[51]

In 2003 PEER conducted a survey of staffers in the EPA's Region 8, which includes most of the states among the administration's leading targets for energy development—Colorado, Utah, Wyoming, Montana, and the Dakotas. The survey found widespread demoralization caused by the political pressure to please industry. Fully one-third of the respondents said they feared retaliation for performing their job, a feeling that was most pronounced among managers. Wes Wilson, an environmental engineer with the EPA's Region 8 and the legislative representative for Local 3607 of the American Federation of Government Employees, agrees. "We're seeing a pattern here—competent people who are doing objective work being moved because they haven't acquiesced to the Bush administration's radical ideology."[52]

The desire to surround itself with compliant scientists has led the administration to engage in "outsourcing," a practice that effectively transfers science from dispassionate professionals to the corporate boardroom. As part of an initiative by the Office of Management and Budget to outsource 425,000 federal jobs, thousands of environmental science positions are being contracted out to industry consultants already in the habit of massaging data to support corporate profit taking. This program would place the care and oversight of a vast range of public policy and public resources in the hands of people without civil-service or whistle-blower protection—and with little or no duty to the public good.[53] If we continue to squander decades of valuable technical expertise and rely on industry data and risk assessments, Wilson says, "the nation runs the risk of not having a government capable of analyzing its health risks and environmental risks."[54] That means legislators making decisions that will profoundly affect our lives and our children's lives with no better guide than the profit motive.

A case in point: In November 2003 the administration was forced to respond to a lawsuit by the NRDC, the United Farmworkers of America, and other environmental groups against the EPA over its failure to regulate certain pesticides, notably atrazine. Atrazine was first approved in 1958, and by the 1980s it had been implicated by epidemiological studies in a host of illnesses, including prostate cancer and infertility. The data was so compelling, in fact, that the pesticide was banned in the European Union. Despite this, it is still the most heavily used weed killer in the United States. Testing by the U.S. Geological Survey regularly finds alarming concentrations of atrazine in drinking water across the corn belt. In 2002 scientists at the University of California Berkeley found that atrazine at one-thirtieth the government's supposedly "safe" level causes grotesque deformities in frogs, including multiple sets of organs.[55] Last year epidemiologists from the University of Missouri found that atrazine may lead to reproductive abnormalities in humans, including sperm counts that are 50 percent below normal.[56]

In September 2001, as a result of NRDC's lawsuit, the courts agreed that the administration must review the health effects of atrazine. But instead of handing the job to the EPA, the administration asked Syngenta, a Swiss company with U.S. headquarters in Greensboro, North Carolina, to monitor the environmental effects of the pesticide. And just what does Syngenta manufacture? Atrazine! It's almost funny. In an interview with the *Los Angeles Times,* Sherry Ford, a Syngenta spokesperson, said without apparent irony, "This is one way we can ensure that it's not presenting any risk to the environment."[57]

A similar bargain in Florida illustrates the kind of fantastic results we might expect when government relies on data from industry biostitutes. In July 2003, the EPA's regional office overseeing the Everglades approved an absurd study performed by a developer-financed consultant concluding that wetlands discharge more pollutants than they absorb. Bush administration regulators are now using that study to justify giving developers credit for improving water quality by replacing natural wetlands with golf courses and other developments.[58] The findings, of course, contradict everything known about wetlands, including a comprehensive study by the Tampa Bay National Estuary Program, in which more than 25 scientists determined that south Florida wetlands remove far more nitrogen pollutants than they generate.[59] Bruce Boler, a biologist and water quality specialist working for the office, resigned in protest. "It was like the politics trumped the science," he says. "I felt like I was bushwhacked."[60]

During the first few years of Bush's presidency, the assault on science was still somewhat ad hoc. But by midterm, his advisers were moving to institutionalize the corruption. And no one was more zealous in this regard than John Graham.

In September 2003, Graham attempted a massive structural change in how science is generated and how the public gets to hear about it. In what amounted to a brazen coup d'état, Graham has proposed that the Office of Management and Budget take over the peer-review process for all major government rules, plans, proposed regulations, and pronouncements. Under the current system, individual federal agencies typically invite outside experts to review the accuracy of their science, whether it is about the health effects of diesel exhaust, industry injury rates, or the dangers of eating beef that has been mechanically scraped from the spinal cords of mad cows. Graham's proposal would block all new federal regulations until the underlying science passes muster in a peer-review process centralized in his office.[61]

In other words, all government research would be under the control of the guru of junk science. Karl Kelsey of the Harvard School of Public Health recalls reviewing Graham's résumé during his tenure process: "He had this separate tiny paragraph or two about peer-reviewed reports. He's a man who doesn't know what peer review is. It's a horrible thing."[62]

Certainly, placing peer review under OIRA is the ultimate delay tactic. No scientific study is airtight. Science is all about divining conclusions from the data. No matter what the study, no matter who's designed it, Graham's peer-review panels could always find a hole in the data. Remember how the tobacco industry hid behind "scientific uncertainty" to shield cigarettes from regulation for 60 years? Using the same methods, the coal industry, with even greater profits at stake, could shield itself from regulations for acid rain, mercury contamination, and global warming—maybe forever.

Lined up in support of Graham's proposal are all his old friends from the Harvard center, the same industrial polluters who funded the Bush campaign. Opposing Graham are some of the nation's most respected scientific bodies: the National Academy of Sciences, the American Association for the Advancement of Science, the Federation of American Scientists, and the Association of American Medical Colleges, along with environmental and consumer-interest groups. Additional criticism has come from a group of 20 former federal officials, including prominent regulators from the administrations of Richard Nixon, Gerald Ford, Jimmy Carter, George H. W. Bush, and Bill Clinton.

Graham's proposal is now working its way through the regulatory process, and a version of it will almost certainly become law of the land if Bush is reelected in November.

Science, like theology, reveals transcendent truths about a changing world. The best scientists are moral individuals whose business is to seek the truth. Corruption of this process undermines not just democracy but civilization itself. The

Union of Concerned Scientists, a nonprofit group devoted to the use of sound science in environmental policy, issued a report in February 2004 entitled "Scientific Integrity in Policymaking: An Investigation of the Bush Administration's Misuse of Science."[63] The report, signed by some 60 renowned scientists, including 20 Nobel laureates, made news for about 20 minutes in the mainstream media. But its charges were scalding and astonishingly direct. "There is strong documentation of a wide-ranging effort to manipulate the government's scientific advisory system to prevent the appearance of advice that might run counter to the administration's political agenda," the authors wrote. "There is significant evidence that the scope and scale of the manipulation, suppression, and misrepresentation of science by the Bush administration is unprecedented." That's not just some protester talking. Those are the conclusions of our country's top researchers—America's brain trust.

In February 2004, Michael Oppenheimer stated the situation plainly. "If you believe in a rational universe, in enlightenment, in knowledge, and in a search for the truth," he said, "this White House is an absolute disaster."[64]

6

Blueprint for Plunder

There is no better example of the corporate cronyism now hijacking American democracy than the White House's cozy relationship with the energy industry. In the 2000 election, it contributed more than $48.3 million to George W. Bush and the Republican Party, and it has donated another $58 million since the president's inauguration.[1] The investment matured almost immediately: It may be hard to find anyone on Bush's staff who doesn't have extensive corporate connections, but fossil fuel executives rule the roost. Both Bush and Cheney came out of the oil patch. Thirty-one of the Bush transition team's 48 members had energy-industry ties. Bush's cabinet and White House staff is an energy-industry dream team—4 cabinet secretaries, the 6 most powerful White House officials, and more than 20 high-level appointees are alumni of the industry and its allies. It's a tight-knit group: Chevron named an oil tanker after National Security Advisor Condoleezza Rice, who served for a decade on the company's board.[2]

Few of these appointees seem to be troubled about conflicts of interest. In early 2001, while the administration was formulating its national energy policy, Karl Rove still owned $250,000 in Enron stock and tens of thousands more in BP Amoco and Royal Dutch Shell. When asked if Rove ever recused himself from policy discussions involving these companies, the White House stated, "He's involved in virtually every decision that is made here." Clay Johnson, as director of presidential personnel, held $100,000 to $250,000 in stock in El Paso Energy Partners, a Houston oil and gas company. As part of his White House duties, he was involved in selecting people to fill vacancies at the Federal Energy Regulatory Commission, which oversees the natural-gas market. Lewis Libby, Cheney's chief of staff, is a lawyer who in early 2001 sold tens of thousands of dollars' worth of stocks in Enron, ExxonMobil, Texaco, and Chesapeake Energy. Deputy Commerce Secretary Samuel Bodman was the CEO of Cabot, a Texas chemical company cited as the fourth-largest source of toxic emissions in the state. Kathleen Cooper, undersecretary of commerce for economic affairs, was chief economist at ExxonMobil.[3]

The fossil fuel crew was among the first to benefit from the Inauguration Day freeze on environmental regulations. Many of Graham's early moves to emasculate the country's environmental laws seem motivated by his desire to fill up the coffers of the oil, gas, mining, and other utility companies. No industry gained more from the administration's massaged data. But the first wave of looting was nothing compared to the spoils to come from rewriting the nation's energy policy.

Days after his inauguration, President Bush launched the National Energy Policy Development Group, chaired by Dick Cheney.[4] Commonly known as the energy task force, the group was convened ostensibly to analyze America's energy needs and to develop recommendations for meeting those needs. But it behaved more like a band of pirates divvying up the booty.

Cheney was determined to operate his task force in total secrecy. His problem was the Federal Advisory Committee Act (FACA), which requires that the activities of groups that combine governmental and nongovernmental officials be fully disclosed to the public. Cheney decided to make an end run around FACA by limiting it to government officials.[5]

Other than being on Bush's staff, the only qualification for membership on the task force seemed to be an energy-industry pedigree. Energy Secretary Spencer Abraham, a former one-term senator from Michigan who received $700,000 from the auto industry in his losing 2000 campaign, was Cheney's number two.[6] He would lead the fight to kill auto fuel efficiency. Joe Allbaugh, director of the Federal Energy Regulatory Commission, was a member of Bush's "iron triangle" of trusted Texas cohorts.[7] Allbaugh's wife, Diane, was an energy-industry lobbyist for three firms—Reliant Energy, Entergy, and TXU—each of which paid her $20,000 during the last three months of 2000, just prior to the start of the task force's deliberation.[8] Joe Allbaugh participated in task force meetings on issues directly affecting those companies, including debates over environmental rules for power plants and—his wife's specialty—electricity deregulation.

Other task force members included Commerce Secretary Don Evans, an old friend of the president's from their early days in the oil business and former CEO of Tom Brown Inc., a Denver oil and gas company; Interior Secretary Gale Norton, who received more than a third of the $800,000 raised for her Senate campaign from the energy industry;[9] and Treasury Secretary Paul O'Neill, former CEO of Alcoa.[10] Recognizing that aluminum-industry profits are directly tied to energy prices,

O'Neill promised to immediately sell his extensive stock holdings in his former company (worth more than $100 million) to avoid conflicts of interest, but despite his reputation for integrity, he delayed the sale until after the energy plan was released. By then, thanks partly to the administration's energy policies, Alcoa's stock had risen 30 percent.[11]

For three months, despite insistent protests from the press and environmental groups, the task force held closed-door meetings. Practically the only outsiders invited to share their views on energy policy were energy-industry representatives. Exactly who, of course, was a mystery to everyone outside the proceedings, as the task force refused to disclose names.

Finally, on May 17, 2001, the task force released the fruits of its labor, the Report of the National Energy Policy Development Group, which became the basis for the Republican-sponsored energy bill. The report was an orgy of industry plunder, transferring billions of dollars of public wealth to the oil, coal, and nuclear industries, which were already swimming in record revenues. (A few weeks earlier ExxonMobil had announced record profits, with earnings up 50 percent from a year earlier.[12]) Paying lip service to conservation and environmental concerns, the report focused almost exclusively on deregulation, giant subsidies, and tax breaks that would benefit virtually every major polluter in the energy industry.[13]

For the first time in history, the nonpartisan General Accounting Office sued the executive branch, demanding access to the records of the task force.[14] (A recent Bush appointee, federal judge John Bates, dismissed the case,[15] and the GAO elected not to challenge the ruling.[16]) Judicial Watch and the Sierra Club prevailed in another suit against Cheney under FACA.[17] The U.S. Supreme Court accepted the case for review in December 2003.[18] Three weeks later, Supreme Court Justice

Antonin Scalia accepted a ride from Cheney on Air Force Two to a duck-hunting outing in Louisiana hosted by Diamond Services Corporation, an oil services company and major Republican Party donor.[19] Scalia then famously refused to recuse himself from the case when the Supreme Court heard arguments in April, despite the clamor to do so by the plaintiffs, the Senate's Democratic leadership, and editorial boards across the country, including those of the *New York Times*, the *Washington Post*, and even the conservative *Salt Lake Tribune*, which wrote, "Scalia's prickly insistence that no reasonable person could question his impartiality in the matter suggests that the rarified air of the Supreme Court has addled the justice's faculties somewhat."[20] In June 2004 the Supreme Court issued its decision, effectively foreclosing the release of the pages prior to the November election. Scalia's concurring opinion urged dismissal of the case based upon executive privilege.

At the same time that the Sierra Club and Judicial Watch had filed their FACA case, the NRDC had also filed a Freedom of Information Act request with the Department of Energy, and when the department did not respond, we sued the DOE. On February 21, 2002, U.S. District Judge Gladys Kessler ordered Energy Secretary Spencer Abraham and other agency officials to turn over records relating to their participation in the work of the energy task force.[21] Under this court order, the NRDC has obtained some 20,000 documents. Although the pages are heavily censored and none of the logs from Vice President Cheney's meetings are included, the documents still allow glimpses into the sausage making.

Cheney, who had run Halliburton for five years in the 1990s, regarded himself an energy expert and considered public hearings or input from opposing factions mere interference.[22] Paul O'Neill, who discussed his dissatisfaction with the energy task force in Ron Suskind's *The Price of Loyalty,* told me he was initially optimistic that he could have a positive influence on the energy debate. But he quickly realized that sensible energy policy was not the committee's agenda.

At its first official meeting, according to Suskind, the task force briefly discussed the California energy crisis and then Cheney took the helm, making clear that his first priorities were tax breaks and deregulation—both things that would enrich the energy industry at public expense without necessarily creating more energy. Cheney's persistent theme was that the nation was facing catastrophic energy shortages, and that these so-called shortages could be used to justify billions in corporate subsidies to his energy cronies and the scuttling of health and environmental safeguards. Experts at the time were scoffing at the notion of an alleged energy crisis, but as Californians had learned, the perception of shortages, even when false, creates a huge bonanza for industry.

"I was against tax incentives that eventually were steamrolled into the task force recommendation," O'Neill told me. "Good business people don't do things based upon tax code inducements. Tax cuts don't increase industry's appetite for more research or development, but they transfer the costs of what they are already doing to the public."[23]

Cheney wasn't terribly interested in what O'Neill had to say. Instead, he opened his door to industry titans, soliciting a wish list of recommendations from lobbying associations such as the American Petroleum Institute, the American Gas Association, the National Mining Association, and the Edison Electric Institute.[24]

To lobby for congressional passage of the Cheney energy plan, more than 400 industry groups enlisted in the Alliance for Energy and Economic Growth, a coalition created by oil, mining, and nuclear interests and guided by the White House.[25] It cost $5,000 to join, "a very low price," Wayne Valis, a lobbyist for the Nuclear Energy Institute, wrote in a fundraising letter.[26] The prerequisite for joining, he noted, was that members "must agree to support the Bush energy proposal in its entirety and not lobby for changes." Within two months, members had contributed more than $1 million. The price for disloyalty was expulsion from the coalition and possible reprisals by the administration, according to Valis. "I have been advised," he wrote, "that this White House will have a long memory."[27]

Task force members began each meeting with industry lobbyists by announcing that the session was off the record and that participants were to share no documents. A National Mining Association official told reporters that the industry managed to control the energy plan by keeping the process secret. "We've probably had as much input as anybody else in town," he said. "I have to take my hat off to them—they've been able to keep a lid on it."[28]

Through the winter and spring of 2001, executives and lobbyists from the oil, coal, electric-utility, and nuclear power industries tramped in and out of the cabinet room and Cheney's office. Many of the lobbyists had just left posts inside Bush's presidential campaign to work for companies that had donated lavishly to that effort. The National Mining Association, which contributed $575,496 to Republicans, had at least nine contacts with the task force, as did Westinghouse, which contributed $65,060. The American Gas Association, which contributed $480,478 to Republicans from 1999 to 2002, had at least eight contacts, as did CMS Energy, which contributed $357,715. The American Petroleum Institute, which contributed $44,301 to Republicans from 1999 to 2002, had contact with the task force at least six times, as did Exelon Corporation, which contributed $910,886.[29]

Executives from Enron, which contributed $2.5 million to the GOP from 1999 to 2002, had contact with the task force at least 10 times, including six face-to-face meetings between top officials and Cheney.[30] After one meeting with Enron CEO Kenneth Lay, Cheney dismissed the request of California's Governor Gray Davis to cap the state's energy prices. According to evidence obtained by Congressman Henry Waxman of California, the task force "considered and abandoned plans to address California's energy problems in its report."[31] Davis told me he'd met with Bush three times and with Spencer Abraham "many, many times" during that period to implore them to exercise federal authority to cap California energy prices. Bush refused, telling Davis that he believed in free markets.[32] That denial would enrich companies like Enron and Diane Allbaugh's client, Reliant Energy Services (which has since been indicted for orchestrating the California energy scam) and nearly bankrupt California.

Energy companies that had not ponied up remained under pressure to give to Republicans. When Westar Energy's chief executive was indicted for fraud in December 2003, investigators found an e-mail written by Westar executives describing solicitations by Republican politicians for a political action committee controlled by Tom DeLay as the price for a "seat at the table" with the task force.[33]

When it was suggested that access to the administration was for sale, Cheney hardly apologized. "Just because somebody makes a campaign contribution doesn't mean that they should be denied the opportunity to express their view to government officials," he said in an interview with the Associated Press.[34]

Cheney's ironfisted control of information extended to his relationship with the president. In one of the most frightening passages in Suskind's book, O'Neill describes how Cheney directed testimony by cabinet officials to persuade President Bush of a looming national energy crisis that would justify giant tax breaks for big oil and big coal, new subsidies for the nuclear industry, and relief from environmental regulations for everyone. At the meeting, each cabinet official made a prearranged statement cleared by Cheney's office and intended to add to the drumbeat of urgency. Cabinet officials were forbidden from engaging in free-ranging discussion in front of the president and were told that he would not read their reports or memos.

According to O'Neill, the president accepted Cheney's recommendations without apparent curiosity or question. By the end of the meeting, having persuaded the president that the nation was facing a crisis, Cheney had a blank check for the obscene subsidies and deregulation that would be his gift to the energy industry.

Of course, although they met with hundreds of industry officials, Cheney and Abraham refused to meet with any environmental groups. But Cheney did make one exception to his policy of secrecy. On May 15, 2001, the day before the task force sent its plan to the president, CEOs from wind, solar, and geothermal energy companies were granted a short meeting with Cheney. Afterward, they were led into the Rose Garden for a press conference and a photo op.[35]

Comparing the final report with documents obtained by the NRDC, it's clear that the big energy companies all but held the pencil as the task force crafted its report.[36] The plan included several provisions authored by Chevron, including one that would make it much easier for certain companies to get EPA permits. A March 20, 2001, e-mail from the American Petroleum Institute to task force staffers contained a draft for an executive order that would require agencies to weaken environmental safeguards that might impact energy supplies, distribution, or use. Two months later, President Bush issued Executive Order 13211, which adopts the American Petroleum Institute's recommendations verbatim.

Cheney's task force also had at least 19 contacts with officials from the nuclear energy industry—whose trade association, the Nuclear Energy Institute, donated $100,000 to the Bush inauguration gala and $437,000 to Republicans from 1999 to 2002.[37] Its payback? The report recommended loosening environmental controls on the industry, reducing public participation in the siting of nuclear plants, and adding billions of dollars in subsidies for the nuclear industry.[38]

Robert Allison, the chairman of Anadarko Petroleum, and the fourteenth biggest Republican Party donor, met directly with Cheney on February 8, 2001, to petition for expanded oil and gas exploration and production on federal lands.[39] His proposal made it into the final energy plan.

In a field of ferocious corporate advocates, Southern Company was among the most adept. The company, which contributed $1.6 million to Republicans from 1999 to 2002, met with Cheney's task force seven times.[40] Faced with a series of EPA prosecutions at power plants violating so-called New Source Review (NSR) standards, the company retained Haley Barbour, former Republican National Committee chairman and now governor of Mississippi,[41] and Marc Racicot, current Republican National Committee chairman and former governor of Montana, to lobby the administration on its behalf.[42] Chief among their requests was avoiding limits on carbon dioxide pollution from power plants and the gutting of the

NSR rule. This rule, which most environmentalists consider the heart and soul of the Clean Air Act, requires the 1,500 dinosaur power plants that were grandfathered under the Clean Air Act to install modern pollution equipment whenever they upgrade or expand. Despite the fact that every year thousands of citizens in each state die prematurely of respiratory illnesses that can be linked to polluted air from grandfathered plants, several companies had tried to dodge the NSR requirement and were prosecuted criminally by the Clinton Justice Department; Southern owned 8 of the 51 power plants targeted by prosecutors.[43]

Barbour and Racicot repeatedly conferred with Abraham and Cheney. They were joined by Edison Electric Institute, whose director, Tom Kuhn, a Bush Pioneer, is a former college roommate, fundraiser, and close personal friend of the president. Edison, the electric industry's major lobbying arm, contributed $598,169 to Republicans between 1999 and 2002,[44] and its members have given $19.7 million to Republicans since 1998.[45]

The White House forced the Justice Department to drop the prosecutions. Justice lawyers were "astounded" that the administration would interfere in a law enforcement matter that is "supposed to be out of bounds from politics."[46] The EPA's chief enforcement officer, Eric Schaeffer, resigned. "With the Bush administration, whether or not environmental laws are enforced depends on who you know," Schaeffer told me. "If you've got a good lobbyist, you can just buy your way out of trouble."[47]

Among the report's recommendations were exempting old power plants from Clean Air Act compliance and adopting Barbour and Racicot's arguments about NSR and carbon dioxide restrictions. Barbour repaid the favor that week by raising

$250,000 at a May 21 GOP gala honoring Bush; Southern donated $150,000 to the effort.

Cheney wasn't embarrassed to reward his old cronies at Halliburton, either. The company donated $1,030,062 to Republican candidates between 1997 and 2002, and gave all of its soft-money contributions ($535,660) to the party.[48] The final draft of the task force report praises a gas recovery technique controlled by Halliburton. The technique, used in coalbed methane drilling, has been linked to the contamination of aquifers and is currently being investigated by the EPA. A discussion of the human health and environmental risks associated with the process had appeared in earlier drafts.[49] Somehow, that got edited out of the final report.

O'Neill watched the task force proceedings with increasing dismay as incentives to pollute were piled onto the list of recommendations. "Industry has the ability to fix these things, and it does not cost more money," he told me. "It takes energy and technology and leadership. Pollution is a leadership failure."[50]

Conspicuously missing from the task force report is a thoughtful discussion of conservation. "Conservation may be a sign of personal virtue," Cheney explained in a Toronto speech a month before the report was made public, "but it should not be the basis of comprehensive energy policy."[51]

The vice president has his blinders on. Conservation is indeed the fastest way to reduce our dependency on foreign oil. Since 40 percent of the oil used by the United States fuels light trucks and cars, making our vehicles more fuel-efficient is the smartest energy investment. A 1-mile-per-gallon improvement would yield double the oil that could ever be extracted from the Arctic National Wildlife Refuge—and would do it without destroying the country's last great wilderness.[52] A 2.6-mpg improvement would produce more oil than Iraq and Kuwait imports combined. An 8-mpg increase would eliminate the need for all Persian Gulf imports.[53] The $20 billion that we would no longer have to send to the Middle East would help balance our trade deficit and provide a permanent economic stimulus package. With every American pocketing hundreds of dollars in annual savings, conservation would produce more oil per dollar spent and create far more American jobs than drilling in the Arctic and Saudi Arabia.

I drive a minivan that gets 22 miles per gallon and spend $2,300 for gasoline each year. A 40-mile-per-gallon car would leave $1,000 in my pocket every year. Remember when President Bush sent us each a $300 check and called it an economic stimulus package? What would it mean for economic stimulus if we were all getting hundreds of dollars every year in fuel savings—and all without gutting the Social Security Trust Fund?

In 1979, President Carter implemented "corporate average fuel economy" or CAFE standards that encourage carmakers to make more fuel-efficient cars. After CAFE, fuel economy rose 7.5 miles per gallon and helped turn an oil shortage into a glut.[54] In 1986 President Reagan rolled back those standards as a favor to big oil and Detroit.[55] According to a recent report by economist Amory Lovins of the Rocky Mountain Institute, the nonprofit energy research outfit, if the United States had continued to conserve oil at the rate it did from 1979 to 1985, it would no longer have needed Persian Gulf oil after 1986.[56] Had we continued this wise course, we might not have had to fight the Persian Gulf war, and we would have insulated ourselves from price shocks in the international oil market. Fuel efficiency is sound national energy policy, economic policy, and foreign policy all wrapped into one.

The United States, which uses 25 percent of world's oil and sits on 3 percent of global reserves (compared with the Persian Gulf states' 65 percent), can never drill its way out of dependence on foreign oil.[57] Even the conservative Cato Institute called the Bush-Cheney claim that Arctic oil would reduce gas prices and dependency on foreign oil "not just nonsense, but nonsense on stilts."[58]

Of course, the one thing conservation does not do is create massive profits for corporations. CAFE standards force car manufacturers to put money into R&D to come up with better-designed cars; oil companies, for their part, sell less product. Since 1990, $80 million of checkbook diplomacy between the automobile industry and Washington has dulled America's political commitment to fuel efficiency and bought Detroit political connections that rival those of big oil.[59] Now Dick Cheney would rather let oil companies plunder our natural heritage than make auto companies clean up the world's most inefficient gas guzzlers.

As Cheney pushed for passage of his task force recommendations in Congress, Republicans simultaneously refused to renew the tax deduction that had encouraged Americans to buy gas-saving hybrid cars, while the Bush administration weakened efficiency standards for everything from air conditioners to automobiles. They also created an obscene $100,000 tax break for Hummers and the 38 other biggest gas guzzlers. As a result, the United States has its worst energy efficiency in 20 years.[60]

If the president is serious about ensuring our national and economic security, we should immediately set a course to raise fuel economy standards to 40 miles a gallon by 2012, and 55 by 2020. This would give automakers ample time to adjust their production. He should also close the sport-utility vehicle loophole by holding SUVs and minivans to the same fuel economy standards as cars; automakers already have the technology to achieve this. Along with the other benefits, higher fuel economy standards could bring increased demand for efficient cars and the technologies needed to build them, leading to an increase in motor-vehicle-related jobs.

The Bush-Cheney energy plan will make us more dependent on foreign oil, and it will place our hopes for national energy security in an aging, insecure Arctic pipeline that is a sitting duck for terrorists. There is no reason to wait 10 years for Arctic oil to come on line or 20 years for hydrogen fuel cells—President Bush's only solution to long-term oil dependence—when a relatively small investment in conservation would quickly reduce American demand for oil.

But no one on the task force, and certainly no one consulting with it, had any interest in reducing American demand for oil. Instead, Dick Cheney's task force report promotes increasing consumption. Indeed, Bush views his massive tax cuts as a way of helping Americans pay for inflated energy bills and further enrich his oil-industry chums. "If I had my way," he declared at a May 2001 White House press conference, "I'd have [the tax cuts] in place tomorrow so that people would have money in their pockets to deal with high energy prices."[61]

For two years after the energy task force report was released, Senator Tom Daschle and a Democrat-controlled Senate fought to block the Cheney energy plan in Congress. While Daschle and his colleagues were wrangling, however, the White House managed a historic end run around the democratic process. Through a variety of stealth tactics, the administration and its corporate toadies within the federal agencies implemented most of the plan's more grievous provisions. These under-the-radar moves were so successful that Energy Secretary Abraham, in a speech to the U.S. Energy Association in June 2002, reported that the administration had already put into effect three-quarters of the task force's 105 recommendations.[62]

In October 2001, for instance, the Bush White House eliminated a Clinton administration regulation granting veto power to the Interior Department for mining permits that would cause "substantial and irreparable harm" to the environment.[63] The White House later announced it would weaken plans to regulate three major pollutants—mercury, sulfur dioxide, and nitrogen oxide. On August 27, 2002—while most of the country was heading off for the Labor Day weekend—the administration announced that it would redefine air pollution so that carbon dioxide, the primary cause of global warming, would not be subject to regulation under the Clean Air Act.[64] The next day, the White House weakened the Clean Air Act's New Source Review provision.[65] Although the regulation may be reversed in the courts (the NRDC is suing), the damage will have been done and power utilities such as Southern Company will escape criminal prosecution.

One of my lawsuits, a 14-year power plant case, was another casualty. In 1990, several Waterkeepers from across the country sued the EPA to force the agency to stop massive fish kills at power plants. Using antiquated technology, power plants often suck up the entire freshwater volume of large rivers, killing obscene numbers of fish. Just one facility, the Salem nuclear plant in New Jersey, kills more than 300 billion Delaware River fish each year, according to Martin Marietta, the plant's own consultant. Nationally, power plants kill over a trillion fish a year, contributing to the collapse of global fisheries. These fish kills are illegal, and in 2001 we finally won our case. A federal judge ordered the EPA to issue regulations restricting power plant fish kills. But soon after the Cheney task force released its report, John Graham and industry lackeys at the EPA replaced the proposed new rule with clever regulations allowing business to proceed as usual.

The carnage never stopped. By summer 2003, the body of environmental law that had been carefully constructed over the last three decades had become a virtual piñata for energy moguls, delivering new gifts with every blow. In August, the administration proposed limiting the authority of states to object to offshore drilling decisions and ordered federal land managers across the West to ease environmental restrictions for oil and gas drilling in national forests. It also proposed removing federal protections for most American wetlands and streams. "It's almost like they want to alienate people who care about the environment," said one astounded Republican, Congressman Christopher Shays of Connecticut, "as if they believe that this will help them with their core."[66]

Perhaps the most galling concession that Bush made to the energy industry, however, was his announcement that he would refuse to support Superfund. The move went largely unheralded because the president announced the change in his 2003 budget request, disguised as a "reform" of the Superfund program.[67]

It was a decision with catastrophic repercussions. One in four Americans live near a Superfund site, and without sufficient funding, the program is essentially a paper tiger.[68] The Superfund trust historically received much of its money from a tax on 43 particularly hazardous chemicals and a 9.6-cent-per-barrel tax on crude oil.[69] This was a small concession by the oil industry, whose lobbyists had won an exclusion of petroleum products from liability under Superfund's cost-recovery provision. While Superfund can be used to respond to public health emergencies and clean up orphaned sites—where the polluter had gone bankrupt or wasn't known—it is despised by industry not so much because of the small tax but because it constitutes the government's principal lever against recalcitrant polluters. The EPA can use the fund to clean up the site and then bill the responsible polluter triple the cost of cleanup.[70]

Without the Superfund tax, however, the EPA can't do the initial cleanup and has therefore lost the leverage that it needs to force action. The tax expired in 1995 and congressional Republicans refused to reinstate it.[71] In his 2003 budget announcement, Bush followed suit, becoming the first president to abandon the tax, and in October 2003 the fund went bankrupt. Big polluters will save hundreds of billions, and the cost of cleanup will be shifted to the public.

The conspicuous disregard for American citizens that led to the Superfund fiasco is the prevailing pattern of the administration's energy agenda. Of the task force report, O'Neill told me sadly, "The product didn't turn out to be what it ought to have been." When I asked him why Cheney, an intelligent man, could not accept that sound environmental policy is good for our economy, he said, "I don't know what motivates Dick Cheney." When I pressed him as to why the vice president would do things that are so obviously antagonistic to the public good, reminding him that he had known Cheney since they were young, O'Neill said, "Beats the hell out of me."[72]

Maybe not, though a look at the nation's largest energy provider should yield a clue.

7

King Coal

In May 2002 I flew over the hills of West Virginia, Kentucky, and Tennessee, and saw a sight that would sicken most Americans. The mining industry is dismantling the ancient mountains and pristine streams of Appalachia through a form of strip-mining known as mountaintop removal. Mining companies blow off hundreds of feet from the tops of mountains to reach the thin seams of coal beneath. Colossal machines dump the mountaintops into adjacent valleys, destroying forests and communities and burying free-flowing mountain streams in the process. I saw the historic landscapes that gave America some of its most potent cultural legends—the forests where Daniel Boone and Davy Crockett roamed, the hills that bred the soldiers who followed Andrew Jackson, the frontier hollows that cradled our democracy, the wilderness wellspring of our values, our virtues, our national character—all being leveled.

According to the EPA, the waste from mountaintop removal has permanently interred 1,200 miles of Appalachian streams, polluted the region's groundwater and rivers, and rendered 400,000 acres of some of the world's most biologically rich temperate forests into flat, barren wastelands, "limited in topographic relief, devoid of flowing water."[1]

At the current rate, another million acres will disappear within decades. That's a total of 2,200 square miles—an area the size of Delaware.[2]

The EPA's findings only confirm what has long been obvious to the people of Appalachia. "I look at what they're doing and I can see the moonscape that they've created. And it's total devastation, total devastation. Nothing will ever grow back," says Judy Bonds, a 52-year-old grandmother from Whitesville, West Virginia. Bonds runs Coal River Mountain Watch, a community group that opposes mountaintop removal. In 2001, her courageous battle against the coal barons won her a Goldman Environmental Award—the Nobel Prize of community activism. "I tell my grandson when he sees these logging trucks go by with all the beautiful hardwood trees on it, 'Son, take a good look at that because neither you, nor seven generations of your children, will ever, ever see trees like that again in West Virginia.'"[3]

We flew underneath a Dragline in our little Cessna 172, dwarfed by the half-billion-dollar backhoe with a scoop big enough to hold 26 Ford Escorts. Below us I could see a half dozen oversized dump trucks. These absurdly colossal machines—along with the 2,500 tons of explosives detonated each day in West Virginia alone—have nearly dispensed with the need for human labor.[4] And that, indeed, is the point.

There was no environmental issue about which my father cared more passionately than strip-mining. He visited Appalachia in 1968 and told me how the coal companies were using this technique to put miners out of work. In the process, they were also destroying our historic landscapes and permanently impoverishing the region. Strip-mining made its debut in the 1940s in the western states, to get at the coal seams that were just a few feet below the surface and inaccessible through traditional tunnel mining. To extract the wealth all you needed was a bulldozer. In Appalachia, the mining companies adopted the process to get at deep coal seams. It was a labor-saving practice that allowed the mining companies to decimate unions that had championed worker health and safety for generations. Nothing was left behind, my father said—not even the hope that Appalachia's people could someday resurrect their economies or communities. Since my father's trip, the machines and cuts have grown bigger and bigger while the workforce has shrunk. Back then, there were 120,000 coal miners in West Virginia. Today, thanks in part to mountaintop removal, there are fewer than 15,000.[5] To get the same amount of coal as they did in 1960, these companies now use 12 percent of the workers.

Poisoned streams, horrific noise, and choking dust make life near these mines unbearable. "We've watched our communities become ghost towns," says Judy Bonds, whose family has lived in Marfork Hollow for nine generations. Bonds was radicalized in 1997 when she saw her 11-year-old grandson standing in a creek of dead fish poisoned by mining drainage. "We only have one grocery store where we used to have four. And you can walk through the little town and see that most of the buildings are boarded up because the businesses failed and the young people have left the area."[6]

It's the same story wherever King Coal sets up shop. From Appalachia to the western states of Wyoming and Utah, the strip miners have permanently destroyed some of the most beautiful country on the planet, leaving behind a legacy of misery and poverty. King Coal sends more greenhouse gases into the air and more mercury and acid rain onto our earth and produces more lung-searing ozone and particulates than any other industry. As the nation's largest energy provider—more than half of our electricity is coal-fired—big coal is the number one polluter.

It's also a key Bush donor. Coal-mining companies and the coal-burning utilities donated $20 million to President Bush and other Republicans in 2000 and have since sweetened the pot with another $21 million.[7] Their generosity has not gone unnoticed. No industry had more highly placed sympathizers in the Bush camp than King Coal. Lobbyists and executives of coal companies had unparalleled access to Cheney's task force while it was creating its new energy bill. During the two years that the bill was stalled in Congress, coal sympathizers in the White House employed a variety of tactics to push through many of the bill's provisions. At the West Virginia Coal Association's annual conference in May 2002, President William D. Raney reminded the 150 industry moguls in attendance, "You did everything you could to elect a Republican president." Now, he said, "you are already seeing in his actions the payback."[8]

The experience of Peabody Energy, the world's largest coal company, is typical. Executives from Peabody and its Black Beauty subsidiary served on the Bush transition's energy advisory team, and Peabody officials met repeatedly with task force members.[9] When the task force released its final report, it recommended accelerating coal production and spending $2 billion in federal subsidies for research to make coal-fired electricity cleaner.[10] Five days later, Peabody issued a public stock offering, raising $60 million more than analysts had predicted. Company vice president Fred Palmer credited the Bush administration. "I am sure it affected the valuation of the stock," he told the *Los Angeles Times.*[11]

Peabody also wanted to build the largest coal-fired power plant in 30 years upwind of Mammoth Cave National Park in Kentucky, a designated UNESCO World Heritage site and International Biosphere Reserve. With arm twisting from Deputy Interior Secretary Steve Griles and $450,000 in GOP contributions in a three-month period, Peabody got what it wanted.[12] Political appointees with no technical expertise overturned a study warning of air pollution from the plant, and park scientists who expressed fears that several endangered species would be harmed due to mercury and acid rain deposits were ignored. The plant is expected to come on-line in 2007.

The coal barons were apprised of every move by the energy task force, while the rest of the nation was kept in the dark. I recently obtained the transcript of a briefing by Quin Shea, a top lobbyist for the Edison Electric Institute, to a closed-door conference of coal- and utility-industry big shots in April 2001, a month before Vice President Cheney disclosed the administration's energy plan.[13] (The head of EEI, it bears remembering, is Bush's pal and Pioneer donor Tom Kuhn.) Shea had received regular briefings on energy task force business from several White House insiders: task force executive director Andrew Lundquist, chief economic adviser Larry Lindsey, and then–OMB head Mitch Daniels. The transcript of Shea's comments reveal that the Bush administration's energy task force proposals followed a line-by-line game plan devised by his coal and utility contributors.

At the conference, Shea explained that Edison was "working with the vice president" on behalf of the coal industry. Shea refers to the Republican Party as "our party" and the administration as "we." He says: "We desperately want to burn more coal. . . . Coal is our friend." He cautioned, however, that several Clean Air Act and Clean Water Act requirements—in his words, "coal killers"—would soon impose costly cleanup measures on fossil fuel companies unless something was done to scuttle or delay them. Luckily, Shea explained, the administration was coming to the industry's rescue. But he warns his cronies against complacency, telling them that in the future they should not assume that they'll have a president like "Bush or Attila the Hun" who would presumably be as willing to plunder.

Shea boasted that in addition to getting "possible tax relief," they had killed the Kyoto accord: "Kyoto is dead. Kyoto is absolutely dead. . . . For those of you . . . who want to continue to beat that dead horse, let me tell you right now, there will be no equine resurrection here." He noted the Bush administration's desire to abolish New Source Review standards and predicted a reversal of President Bush's campaign promise to regulate CO_2. "We're taking steps right now to reverse every piece of paper that EPA has put together where they could call CO_2 a pollutant under the Clean Air Act," Shea told the assembled executives. "That's going to be worked on in the next few months."

In a telling exchange suggesting that some utility executives do indeed have reptilian hearts, Shea told the group how he explained the Bush rollbacks on ozone and particulates to his own worried grandmother. The EPA's estimates, he noted, predict those pollutants will put 15,000 to 100,000 Americans each year at risk for premature death, mostly children and the elderly.

"Folks," he said, "those are just numbers. They're scary numbers. They scare people. They've scared my grandmother.

She is ninety-seven and said, 'What is going on?' I said, 'Gram, this is wrong. Plus, it's premature mortality. If you die a day early, you're a statistic.' She said, 'Oh, okay.' She didn't really understand, but she sort of got it that I was taking care of it and it wasn't a problem."

Judy Bonds and her friend Freda Williams both live in the shadow of a slurry dam at a mine owned by the Marfork Coal Company, a subsidiary of Massey Energy. Coal dust washed from the pulverized rubble, along with the toxic wastes from mountaintop removal, are stored as a thick, oily sludge in billion-gallon reservoirs behind sometimes shoddily constructed earthen dams.

The Marfork Dam is 925 feet from toe to crest and is capable of holding 9 billion gallons of slurry perched over a deep mine.[14] A breakthrough could bury communities and schools for miles down the hollow.[15] Neither residents nor fire and police departments in nearby Beckley, Madison, or Charleston have ever seen an evacuation plan. "The coal company will tell you, yes, they have a plan, it's on file at their office, and it's not their responsibility to put the word out to the public—to notify the public," says Williams, who has been fighting the dam since 1996 as a member of Coal River Mountain Watch. "If a breakthrough occurs, it will be like an explosion, and there's just not going to be time for an emergency evacuation anyhow. It would take at least an hour for help to get into the area from Charleston."

Everyone in the community lives in constant fear of a breakthrough. Judy Bonds says her grandson "used to lie awake at night when it would rain and plot escape routes in case the sludge dam would bust and drown us, much like Buffalo Creek."

Their fears are not so far-fetched. In 1972, the Buffalo Creek Dam southwest of Charleston collapsed, burying 120 souls and several communities. Just four years ago, on October 11, 2000, a slurry pit in Inez, Kentucky, owned by Martin County Coal, another subsidiary of Massey Energy, burst into subsurface mine shafts, flooding downstream communities. The 300-million-gallon spill was the largest in American history and, according to the EPA, the greatest environmental catastrophe in the history of the eastern United States. Thick black lava-like toxic sludge containing 60 poisonous chemicals choked and sterilized 100 miles of rivers and creeks and poisoned the drinking water in 17 communities. Unlike some other slurry disasters, no one died this time, but hundreds of residents were sickened by contact with contaminated water.

Jack Spadaro was a member of a team of geodesic engineers selected by the Mine Safety and Health Administration, a division of the U.S. Department of Labor, to investigate the spill. Spadaro, the superintendent of the Mine Health and Safety Academy where MSHA trains its engineers, is the nation's leading expert on slurry spills, having spent 30 years studying slurry dam failures and how to prevent them.

From the outset, the coal industry kept a wary eye on the Massey investigation, since it would raise the obvious question of whether slurry impoundments should ever be permitted over abandoned mines. Of the 650 existing impoundments, at least half are in this tenuous situation. Conditions under the surface haven't been investigated for any of them. "I'm worried that the potential for more breakthroughs is great," Spadaro told me. "That question certainly would have been addressed in our report.

"We had a team of engineers who were very effective. We were geotechnical engineers determined to find the truth,"

Spadaro continued. "We simply wanted to get to the heart of the matter—find out what happened, and why—and to prevent it from happening again."[16]

During Spadaro's investigation, however, there was a regime change at the White House. And it became clear that George W. Bush and his coal cronies were just as concerned about the Inez disaster—for very different reasons. Spadaro soon found that all the hard work by his team "was thwarted at the top of the agency by Bush appointees who obstructed professionals trying to do their jobs."

Those Bush appointees all had coal-industry pedigrees. The new Bush team at the Department of Labor included Secretary Elaine Chao, a former fellow of the Heritage Foundation,[17] who is the wife of Senator Mitch McConnell of Kentucky, the Senate's largest recipient of coal-industry largesse. As the investigation moved forward, Massey Energy contributed $100,000 to a Republican Senate campaign committee controlled by McConnell. Chao appointed Dave Laurisky, a former executive with Energy West Mining, a Utah coal company, to be director of MSHA. Laurisky's deputy assistant secretary was John Caylor, an alumnus of Amax Mining; his other deputy assistant, John Cornell, had worked for both Amax and Peabody Coal. Together, this group wasted no time in putting the brakes on the investigation.

Tony Oppegard, Spadaro's boss, whom the team regarded as a strong leader with unquestioned integrity, was fired on the day of Bush's inauguration. "He was getting down to the root of what was going on," Spadaro says. "He was simply fired. The people at Massey knew before he did." Spadaro recalls how Oppegard's replacement, Tim Thompson, laid things out at the first meeting he attended: "We are going to terminate this investigation and make sure that no fingers are pointing at persons within this agency."[18]

The original team had intended to charge Massey with knowing and willful violations of the law. The company had ignored safety recommendations arising from two previous slurry spills; these recommendations came from both the MSHA and its own engineering firm. Massey's engineer had testified following a 1994 accident that if the recommendations were ignored, a new spill was "virtually inevitable." But Thompson reduced eight citations for criminal negligence to two, choosing the weakest—and one of those was later thrown out by the administrative judge. "Massey paid a pathetic $5,600 fine for doing billions of dollars in damage," said Spadaro.

All eight members of the team were pressured to sign off on the whitewashed investigation report. Ronald Brock, one of the team's engineers, refused until Laurisky directly ordered him to sign, according to Spadaro. "Ron Brock's wife was suffering from cancer and he felt he could not afford to lose his job," said Spadaro. "They forced it down his throat."

Spadaro flat-out refused to sign the report and has been harassed ever since. At one point, he says, department officials lured him to Washington for a meeting, and while he was there three men raided his office, changed his locks, searched his papers, and disassembled frames holding pictures of his wife and daughter. "They Gestapoed me," said Spadaro. "I guess they thought I had secret codes written on the back of those photos."

In February 2004 the federal government's independent Office of Special Counsel began an investigation into whether Spadaro was being disciplined because he is a whistle-blower.[19] Less than a week later the MSHA demoted him and reassigned him to a job in Pittsburgh.[20]

"I've been regulating mining since 1966," Spadaro told

me, "and this is the most lawless administration I've encountered. They have no regard for protecting miners or the people in mining communities. They are without scruples. I know that Massey Energy influenced Bush appointees to alter the outcome of our report! The corruption and lawlessness goes right to the top."

In addition to Chao, McConnell, and Laurisky, big coal has placed a legion of guardian angels in the Bush government, including Jeffrey Holmstead, the EPA's assistant administrator for the Office of Air and Radiation. Like many people in the current administration, including George Bush himself, Holmstead can be personable and charming, and like the others, he has elected to spend his talents shilling for industry.

Holmstead has been a lobbyist and attorney for several big polluters and their trade associations,[21] including the American Farm Bureau Federation and the Alliance for Constructive Air Policy, a front group for big air polluters, four of which were defendants in major Clean Air Act enforcement actions for their violations of the New Source Review standards. He has also served as an adjunct scholar with the Wise Use group Citizens For the Environment. CFE pitches clear-cut logging as good timber management and deregulation as the solution to most environmental problems. CFE took $700,000 from the Florida sugar industry to help derail Everglades restoration and received $1 million from Philip Morris to fight cigarette taxes.[22]

Holmstead maintained close personal and professional ties with his former clients that blurred the line between government service and venal self-interest. Within days of his appointment, his buddy David Rivkin landed a job at the law firm of Baker & Hostetler as a lobbyist for Atlanta-based

Southern Company,[23] the nation's number two coal plant polluter[24] and the utility that led the charge to kill the New Source Review rules, which require grandfathered coal plants to install modern pollution-control equipment. One day after Holmstead accomplished that mission, his associate administrator for congressional affairs, Ed Krenik, took a lucrative job at Bracewell & Patterson, the Houston-based lobbying firm that led Southern Company's efforts against the NSR. Holmstead's chief of staff, John Pemberton, left that week to become Southern's top Washington lobbyist.[25]

Holmstead has been dogged by accusations of ethics violations throughout his EPA career. On July 14, 2003, Senator John Edwards of North Carolina called for Holmstead to resign for suppressing scientific research that conflicted with President Bush's pollution agenda and for routinely lying to hype Bush's efforts to roll back environmental regulations. "Jeff Holmstead is an extreme example of this administration's problem with telling the truth when it conflicts with its political agenda," Senator Edwards said. "Instead of protecting the air, Mr. Holmstead is protecting the energy industry by hiding the truth. He needs to go."[26]

An October 2003 report by the GAO indicated that Holmstead had intentionally deceived two Senate committees in an appearance 15 months earlier. At issue was his claim that the administration's proposed scrapping of the NSR rules would not affect 75 ongoing prosecutions and investigations against corporations that had violated those standards.[27] Among the targets of the prosecutions were eight power plants owned by Southern Company's subsidiaries.[28] As it turns out, Holmstead had discussed the impacts on numerous occasions with senior EPA officers. They included Sylvia Lowrance, acting chief of the Office of Enforcement and Compliance Assurance; Eric

Schaeffer, head of the Office of Regulatory Enforcement; and Bruce Buckheit, director of the Air Enforcement Division, all of whom have left the EPA because they say they were disgusted by the administration's pandering to polluters.[29] They and many others had warned Holmstead that their cases would be compromised. "It was clearly not true what Holmstead said," I was told by Buckheit.

The GAO's report prompted Senators Jim Jeffords, Patrick Leahy, and Joseph Lieberman to call for further investigation by the EPA's inspector general.[30] Leahy, the ranking Democrat on the Judiciary Committee, said that the report indicated that Holmstead "intentionally misled Senate committees last year and has continued to work to get industry polluters off the hook and out of court." Leahy accused Holmstead of committing fraud and "endangering the health of the American people."[31]

Endanger our health he has. The EPA's own consultants estimated that air emissions from just 51 of the coal plants that were targets of NSR enforcement actions shorten the lives of at least 5,500 people per year. In addition, polluted air from these plants triggers between 107,000 and 170,000 asthma attacks per year, many of them in children.[32] In 2002, Edwards asked Holmstead for a quantitative study of the NSR proposal's effect on human health. He and 43 Senate colleagues followed up the request in writing. Holmstead has never provided the analysis.[33]

The NSR rollback was only one handout to King Coal that came courtesy of Holmstead. In November 2003, former Utah governor Mike Leavitt replaced Christie Whitman as EPA administrator. Leavitt made his bones with his new bosses in one of his first official acts: proposing weaker regulations for the mercury spewing out of power plant smokestacks. He gave a billion-dollar favor to King Coal and the utilities, while

America's children got the back of his hand. Jeff Holmstead was the deal's architect.

Mercury is a potent brain poison. Even minuscule amounts can cause permanent IQ loss, along with blindness and possible autism in children who are exposed while in the womb.[34] As noted earlier, the EPA now estimates that 1 of every 6 American women carries unsafe levels of mercury in her blood, putting 630,000 American newborns a year at risk.[35] High exposure in adults can lead to kidney failure, tremors, heart disease, severe liver damage, and even death.

I recently had my own blood tested for mercury. The sample came back with contamination of 11 micrograms per liter. Levels above 10 micrograms are cause for concern, according to Dr. David Carpenter of the Institute for Health and the Environment at the State University of New York at Albany. As a healthy adult male, I'm not too worried. But Dr. Carpenter told me that a pregnant woman with those levels would have a child who was cognitively impaired. This made me pause. "You mean 'might'?" I asked him. He replied, "No, the science is pretty certain that those levels would impact a baby's IQ."[36]

Humans get contaminated mainly by eating fish. Ocean-going fish—including favorites like swordfish and tuna—along most of America's coastlines are so contaminated with mercury that they are unsafe to eat regularly. Even more alarming, 17 states have issued warnings about eating fish from *all* of their streams and lakes, including the five Great Lakes and their tributaries. Forty-five states have issued advisories on mercury in fish for some water bodies.[37]

Most mercury in fish comes from contaminated rain and snow that falls into rivers and lakes. According to the EPA, coal-burning power plants account for 40 percent of the airborne mercury in the United States.[38]

Because of the catastrophic health impacts, the Clinton administration decided to regulate mercury emissions from power plants as a hazardous pollutant under the Clean Air Act. That move would have required power plants to reduce their mercury emissions by roughly 90 percent within three years. Technologies now available or expected to be available soon can eliminate most of the mercury from utilities at a cost of less than 1 percent of plant revenues, according to the EPA. This seems like a good deal for the American people.[39]

But the power industry lobbied aggressively against the Clinton-era reforms. The campaign included several Texas companies, whose power plants spew more mercury than those of any other state. Southern Company also lobbied hard. In his briefing to coal moguls in 2001, the EEI's Quin Shea reassured his audience that the coal industry and the Bush administration had a plan that would ensure that no power plant would be required to achieve anything close to a 90 percent reduction in mercury pollution, regardless of what the Clean Air Act said.[40] Their plan was Jeffrey Holmstead.

In December 2003, Leavitt and Holmstead scrapped the Clinton reforms and proposed new ones. These would leave mercury pollution nearly seven times as high as the Clean Air Act would require for at least the next 14 years. But that's only the beginning. Lawyers for these polluters would be able to delay implementation of even those feeble standards using huge loopholes that they themselves buried like land mines in the provision's language. Portions of Holmstead's proposal came verbatim from a memo prepared by Holmstead's old law firm, Latham & Watkins, which represents some of the affected utilities.

After giving his industry cronies rollbacks beyond their wildest dreams, Holmstead continues to troll for new ideas for federal giveaways. In January 2004, he attended a conference for lobbyists at the Arizona Biltmore. The event followed a $3,000-per-person golf-and-dinner Republican fundraiser, dubbed "Mulligans and Margaritas." The conference was touted as an opportunity to draft a to-do list of industry-friendly legislation.[41] Among the top 10 priorities targeted by the conferees were: reforming the Endangered Species Act, the National Environmental Policy Act, and the Clean Air Act; expanding access to public lands; enacting the federal energy bill; and reforming the legal tort system.[42] The gathering was organized by Jim Sims, a Denver-based coal and utility lobbyist and former spokesman for Cheney's energy task force.[43] Also on hand at the boondoggle-fest were James Connaughton, Bush's Council on Environmental Quality director and former asbestos attorney, and deputy secretary of the Interior J. Steven Griles. While Holmstead is an extremely useful ally of energy, the appointment of Steve Griles was arguably the biggest payback to the coal industry.

There was nobody in the United States better positioned than Steve Griles—with his 35 years of working for the energy industry, both in and out of government—to do King Coal's bidding.

Ironically, it was the very statute Griles spent his career trying to destroy that catapulted him to fortune and power. From 1970 to 1981, Griles worked at Virginia's Department of Conservation and Economic Development, which operates a sleepy permit office charged with issuing mining permits to coal companies in Virginia's Appalachian country. During the early 1970s, Griles made a name for himself by running interference on behalf of the coal business. "I found Steve to be extremely pro-industry," recalls Frank Kilgore, a Virginia lawyer who worked on mining reforms during that period. "No matter what evidence you showed him about people having their houses blown apart, or rocks coming through the roof, or private cemeteries or water supplies being destroyed by stripping, it didn't seem to make any impression on him. He was always pretty up-front that he was an industry man—and get out of the way."[44]

Then, in 1977, in the wake of the Buffalo Creek massacre, Congress passed the federal Surface Mining Control and Reclamation Act (SMCRA). Suddenly Griles' agency was required to regulate an industry with which it had always been chummy. He took the new law as a personal affront. He led Virginia in challenging the SMCRA as an unconstitutional infringement on states' rights.[45] Although his suit suffered a rare 9–0 loss in the Supreme Court, it did earn Griles plenty of industry gratitude, the only credential he needed to land a job in Jim Watt's Department of Interior in 1981.

Watt appointed Griles to run the very agency he had vowed to destroy—the Office of Surface Mining (OSM), an agency in the Department of Interior that was created to administer the SMCRA. As deputy director from 1981 to 1983, Griles is said to have promised to "turn the lights out at the OSM."[46] While he neither confirms nor denies the statement, everyone agrees that Griles gutted the agency. He cut staffing by a third, dramatically reduced the number of federal inspectors at mine sites, and sharply curbed enforcement actions.[47] Griles recruited industry-friendly personnel, fired or transferred environmentally friendly regulators, and fought to shift regulatory authority to the states. Staff morale plummeted. Griles himself told the *Washington Post,* "We tore this agency to hell."[48]

Griles continued his campaign as he moved up the food chain in Reagan's Interior Department. When he landed the job of assistant secretary for lands and mineral management at the

Department of the Interior in 1985, the *Oil Daily* endorsed the appointment as the "ideal choice." During his tenure, Griles leased more federal offshore oil and gas acreage than anyone in the history of the Interior Department.[49] He was accused of orchestrating what amounts to a criminally negligent giveaway of 82,000 acres of federal oil and shale lands for a ruinously cheap $2.50 per acre.[50] The government received $200,000. One claimholder got title to 17,000 acres for $42,500—and then immediately sold the same land for $37 million.[51]

Griles tried to open the Arctic National Wildlife Refuge to drilling and vigorously promoted offshore oil leasing in California and Florida. The House Government Operations Committee said Griles' program was so badly administered that Congress should consider transferring it to another agency.[52] In 1989, investigators for the California legislature uncovered internal Interior Department records revealing that Griles deliberately concealed from the public and state regulators the true risks of oil spills from drilling off the California coast.[53]

At the end of the Reagan administration Griles formally went to work for the coal industry. From 1989 to 1995 he was a senior vice president for the Virginia-based United Coal Company, where he oversaw operation of one of the nation's largest mountaintop-removal operations. In 1995 he founded J. Steven Griles and Associates, a lobbying firm that represented over 40 coal, oil, gas, and electric companies and trade associations. Subsequently, he merged his firm with National Environmental Strategies (NES), a lobbying outfit founded by former Republican National Committee chairman Haley Barbour. NES represented the National Mining Association and Dominion Resources, one of the nation's largest power producers, as well as the Edison Electric Institute, Shell, Texaco, Chevron, Arch Coal, and Pittston Coal, to name just a few.

These clients poured millions of dollars into George W. Bush's presidential campaign, and Bush's appointment of Griles as second in command at Interior was exactly what they were paying for. Griles served on President Bush's transition team, helping to fill key administrative posts with reliable lobbyists from regulated industries. Following the announcement of his appointment, the National Mining Association hailed Griles as "an ally of the industry." The *Denver Post* lamented that the "champions of industry will be running the department that oversees most of the nation's public lands."[54]

Like Holmstead, Griles has a tortured relationship with the truth. It's bad enough that a former mining lobbyist was put in charge of overseeing mining on public land. But it turns out that Griles is still on the industry's payroll. In 2001, he sold his client base to his NES partner Marc Himmelstein for $1.1 million, payable in four annual installments of $284,000, making Griles, in effect, a continuing partner with a direct financial stake in the firm's profitability.[55] The Senate made Griles agree in writing that he would avoid contact with his former clients as a condition of his confirmation.[56] Under the agreement, Griles is prohibited from dealing with matters involving NES for six years and is barred from matters concerning "former" clients for a year.

Griles has trouble keeping his promises. His appointment calendar, obtained by Kristen Sykes of the environmental organization Friends of the Earth, indicates that Griles met repeatedly with oil-industry clients to discuss offshore leases in which they had an interest.

One of those companies was Chevron, which paid Griles $80,000 to lobby the Interior Department even while his nomination was pending before the Senate in 2001.[57] As soon as he was confirmed, Griles negotiated a lucrative deal for

Chevron in which the federal government would pay the company a whopping $46 million to drop its plans to drill off the Florida coast. This decision both enriched Griles' former clients and enhanced the reelection prospects of President Bush's brother Jeb as governor of Florida.[58]

Griles continued to flout ethics laws. He had signed a second recusal agreement when he took office, this one banning him from involvement in decisions about a coal-bed methane project in Wyoming and Montana, which was being promoted by six former clients. Career personnel in the EPA's Denver office had delivered a devastating assessment of the proposal, which called for building 51,000 wells in Wyoming's Powder River basin.[59] The project will require 17,000 miles of new roads and 20,000 miles of pipeline and will foul pristine landscapes with trillions of gallons of toxic wastewater.[60] Several months after Griles signed his recusal agreement, he wrote a memo to Linda Fisher, EPA deputy administrator, demanding that she overrule her Denver employees. In his letter, Griles scolded the EPA for investigating the effects of coal-bed methane development on water quality and warned Fisher not to "impede the ability to move forward in a constructive manner."[61] Thanks to Griles, the project was approved.

Griles again violated his recusal pledge by meeting with the National Mining Association, a former client, while the industry group was lobbying the administration to relax restrictions on mountaintop-removal operations. He also had 14 other meetings on mountaintop removal with industry and government officials.[62] And then on April 15, 2002, he assembled the officials who oversee the mining and drilling operations of his former clients for a dinner party at the home of the owner of NES, Marc Himmelstein, who was now representing those clients. Included at Himmelstein's soiree were Rebecca Watson, assistant secretary for Land and Minerals Management; Kathleen Clark, director of the Bureau of Land Management; and Jeffrey Jarrett, director of the Office of Surface Mining.[63]

With Republicans chairing the committees, no congressional subpoenas have interrupted the Griles scandals. In fact, an Interior Department spokesman has gone so far as to say that Griles' conduct represents "the highest ethical and professional standards."[64]

In May 2002, Senator Joseph Lieberman asked the Interior Department to investigate Griles. On June 3, 2003, environmental and government ethics organizations, including the NRDC, joined Whitney North Seymour Jr., former independent counsel and Richard Nixon's U.S. attorney for the Southern District of New York, in calling for the appointment of a special counsel to conduct a criminal investigation of Griles for perjury and ethics laws violations, and for steering government contracts to former clients.[65] The Interior Department has refused to turn over 300 pages of documents concerning NES's $1.1 million payment to Griles that environmental groups requested in September 2002 under the Freedom of Information Act.[66]

On March 16, 2004, the Office of the Inspector General for the Interior Department concluded an 18-month investigation, commenced at Senator Lieberman's request, with a report setting forth strong evidence of unethical conduct by Griles. The report characterized the initial choice to appoint Griles, with all his inherent conflict of interest, as "a train wreck waiting to happen." It confirmed that Griles had regular dealings with energy- and mining-industry clients of his former lobbying firm even as he continued to receive income from the firm's owner. The report concluded that evidence showed that "the department's leadership did not take ethics seriously."[67]

The inspector general describes an ethical atmosphere within Interior so lax that when an "onslaught of public criticism erupted" over Griles' dinner at Himmelstein's, Griles was told by Timothy S. Elliott, the department's deputy associate solicitor, that there would be no problem so long as Griles paid for the dinner himself.[68] Griles wrote Himmelstein a $180 check, which went uncashed for many months.

Investigators complained of the rough-man handling they got from Griles, Norton, and their powerful friends. The investigation, the report says, was obstructed by "an unanticipated lack of personal and institutional memory; conflicting recollections; [and] poor record-keeping." The report added, "When we interviewed the Deputy Secretary and discussed our efforts to discern the status of his client list, he commented simply, 'Good luck.'"[69]

When investigators questioned Griles and Himmelstein about Griles' involvement in the federal payoff to former clients Shell and Chevron, in light of the formal recusal he had signed banning him from any dealings with those companies, Griles told them that he had listed these companies erroneously on his recusal form. Himmelstein claimed, apparently with a straight face, that Griles had not lobbied for Chevron, despite Griles' having been listed as a Chevron lobbyist in filings with ethics offices.

Griles told the inspector general that he could not explain why Chevron's first six payments to his firm included the annotation "Attn.: Steven Griles." At first he maintained that his meetings with Shell's subsidiary, Aera, were permitted because his recusal did not apply to subsidiaries; he later changed tack and claimed he was unaware that Aera was owned by Shell.[70]

In a letter to Senator Lieberman, Inspector General Earl E. Devaney indicated his suspicion that in at least two instances

Griles had violated ethics rules; Devaney also forwarded his report to the Office of Government Ethics for resolution. But that office referred the two cases to Interior Secretary Norton, who was also criticized in the report.[71] Norton announced that she considered the case closed. "I'm glad that we can now put these allegations behind us," she said.[72] Griles added with finality, "I am glad this matter is behind me and we can continue to work to advance our initiatives."[73] Devaney had presciently complained in his transmittal letter to Norton that the American people might never get "a sound legal conclusion" on Griles' activities inside and outside government as well as the widespread ethics abuses in her department.[74]

Griles enjoys a level of access to the White House usually reserved for cabinet officials. During his first 15 months on the job, he attended dozens of White House meetings, including audiences with President Bush and two with Karl Rove, and White House officials have invited him to at least 30 meetings.[75] Bush quickly deployed Griles to help shape national energy policy. He met at least 37 times with industry officials to help craft Bush's Clear Skies agenda, New Source Review standards, and other Clean Air Act rollbacks, even though Interior has almost no jurisdiction over air issues.[76] Griles knew the terrain, having lobbied for 13 industry clients on clean air issues before becoming deputy secretary.

But nothing would earn the coal industry's gratitude more than Griles' efforts on behalf of mountaintop-removal mining. Since the 1970s, Griles has worked to allow this devastating practice to flourish.

Before Bush was elected, Appalachia still had a few powerful legal tools at its disposal that Griles hadn't managed to dismantle during his Reagan-era stint. One of them was the Clean

Water Act, which prohibits the discharge of waste into U.S. waterways without a permit from the EPA. These permits may only be issued if the polluter complies with rigorous standards meant to ensure that there is no decline in water quality. Congress, however, gave limited authority to the Army Corps of Engineers to issue permits allowing the placement of "fill material" in water to build docks, jetties, bulkheads, or other beneficial developments. However, since 1975 the definition of "fill material" had explicitly prohibited fills composed of waste. Nonetheless, the Corps had been permitting coal companies to dump mountaintop-removal waste into streams for years—ostensibly under this limited permitting authority—even though the agency had no legal power to do so. The dumping of mountaintop-removal debris into wilderness streams clearly is not a "beneficial development." In 1998, when Corps official Rodney Woods was asked during a deposition in Cincinnati why the Corps had been illegally approving the disposal, he stated that his agency "just sort of oozed into that."[77]

In 1998, Joe Lovett, an attorney who runs the Appalachian Center for the Economy and the Environment, and Jim Hecker, from Trial Lawyers for Public Justice, sued the Army Corps of Engineers and state regulators on behalf of local citizens and a regional Appalachian group.[78] The law was absolutely clear and the case ultimately resulted in a decision in October 1999 by Chief Judge Charles Haden of the Southern District of West Virginia Federal District Court. Judge Haden declared that it was illegal to dispose of mining waste in streams. He wrote that the practice was a violation of the Clean Water Act's water quality standards, noting that "Valley fills are waste disposal projects so enormous that, rather than the stream assimilating the waste, the waste assimilates the stream."[79]

That decision was the principal reason the industry wanted its guardian angel back in government. On August 5, 2001, three days after signing his recusal letter, Griles brought a reassuring message to a gathering of the West Virginia Coal Association, an audience that included his former clients. "We will fix the federal rules very soon on water and spoil placement," he said.[80] It was a blatant reference to the dumping of waste from mountaintop removal into streams.

Then, in May 2002, on the day I flew across Appalachia, the Bush administration engaged in one of its most cynical maneuvers yet. At Griles' urging, the EPA and the Army Corps of Engineers followed the advice of the National Mining Association and redefined the waste from mountaintop mining as fill. Unfortunately, the prohibition against using waste material as fill, though clearly representing the intent of Congress when the Clean Water Act was passed, did not appear in the act itself.

The Bush rule change created a loophole in the Clean Water Act big enough to drive a Dragline through. And it doesn't just affect Appalachia. Other industries—hard-rock mineral mining, demolition companies, waste disposal operations—all may take advantage of this new loophole. They too may obtain Army Corps permits to bury wetlands, streams, and other waters with their wastes. This is the most significant weakening of the Clean Water Act since it was passed three decades ago.

Chief Judge Haden immediately struck down this new definition *sua sponte*—meaning on his own impulse, without prompting by either side—calling it an "obvious perversity." But the notorious right-wing judges on the Fourth Circuit Court of Appeals upheld the White House's rule change.

Emboldened by this victory, Griles pressed forward to

eliminate the last major federal obstacles to mountaintop removal.

The 1998 lawsuit by Hecker and Lovett contained two other charges against King Coal. First they challenged a permit issued by the Army Corps to Griles' then-client Arch Coal that allowed construction of the largest strip mine in history, despite the fact that the Corps had not performed the environmental impact statement mandated by the National Environmental Policy Act. Lovett and Hecker also charged the Corps with violating the "buffer zone" rule under the Surface Mining Reclamation Act (SMCRA), which forbids the discharge of mining waste within 100 feet of larger streams.[81] Again, both the law and the government's violations were clear as daylight.

Judge Haden, who forbade Arch Coal from constructing the mine, then went a few steps further, forbidding the Corps from issuing any more permits for mining activities in buffer zones, a ruling that would have spelled the end for large mountaintop-removal mines. The conservative Fourth Circuit reversed Judge Haden in a wacky decision that held that the buffer zone challenge should have been brought in state court. Lovett and Hecker refiled in state court in 2003 but moved forward with the EIS challenge in federal court.

Meanwhile, the EPA official responsible for Appalachia, Mike McCabe, recognizing the merit of Lovett and Hecker's position, settled the federal case by agreeing to prepare the EIS and to require greater scrutiny for future permits. As a result of the settlement, no new mountaintop-removal mining permits were issued for approximately two years. With new mountaintop removal temporarily on hold, Lovett and his group began negotiating with the federal government. Several federal agencies—the EPA, the Office of Surface Mining, Fish and Wildlife, and the Army Corps of Engineers—started working with the state of West Virginia on this EIS in 1999. "I thought we were making some progress," recalls Lovett. McCabe said that he did not foresee fills greater than 75 acres ever being allowed in Appalachia after the thorough study of the destruction mandated by the EIS. "The problem," Lovett explains, "is that Bush became president and the whole thing went to hell."

The Bush administration took over the EIS, handing it to Steven Griles to rewrite to suit King Coal.[82] In April and May of 2002, Lovett, Hecker, and their clients got a look at the scientific studies that would form the EIS. They were released to them by a decent and courageous midlevel bureaucrat in the regional EPA office.

The documents contained bad news for King Coal. Even smaller fills, several studies revealed, would permanently destroy vast portions of the unique Appalachian environment.

Griles was undeterred. On October 5, 2001, barely three months after recusing himself from involvement with issues affecting his former coal-industry clients, he sent a letter to five federal agencies saying that the EIS, rather than focusing on minimizing environmental impacts, "must consider and recommend resolutions that will allow steep slope Appalachian coal mining [i.e., mountaintop removal] to proceed." Griles continued: "We do not believe that the EIS, as currently drafted, focuses sufficiently on these goals." He went further: "We must ensure that the EIS lays the groundwork for coordinating our respective regulatory jurisdiction in the most efficient manner," he wrote. "At a minimum, this would require that the EIS focus on centralizing and streamlining coal mine permitting." Griles sent his letter to James Connaughton, White House Council on Environmental Quality director, and Jeffrey Jarrett, Office of Surface Mining, as well as high-level decision makers in the OMB, the EPA, and the Army Corps of Engineers.[83]

Just one month after he sent the memo, coal interests contributed $150,350 in soft money to the Republican National Committee.[84] Griles' former client, Arch Coal, kicked in $15,350 that month, part of $76,894 the company has given to the Bush campaign and the RNC since 1999.[85]

Griles proceeded to rewrite the EIS to weaken safeguards against mountaintop-removal mining. Eight million dollars' worth of compelling scientific and technical studies running over 5,000 pages were moved into appendixes. The main body of the EIS became a discussion about how to make it easier for the coal industry to get permits.

"There's a surreal kind of disconnect between the science of the EIS and the recommendations that came from it," Lovett says in his dry Appalachian twang. "The purpose of the EIS was to study the environmental and social impacts of mountaintop removal and find ways to limit the destruction. Griles ordered them instead to use the process to make it easier for the coal industry to get permits. It's already easy enough for the coal industry to get permits."

Meanwhile, Lovett's case on the buffer zone issue was now close to a decision in the state administrative appeals process, and the coal companies were frantic to derail it. Buffer strips are the last remaining impediment to creating massive fills. Griles decided to use the EIS as a vehicle for jettisoning the buffer zone rule. He inserted in the EIS a proposal to repeal the stream buffer rule, a change that will legalize the obliteration of Appalachia's streams.

Over 80,000 people filed comments specifically opposing the devastating buffer zone rollback, and 12 GOP House members wrote Bush urging him not to make the "ill advised and dangerous" rule change.[86]

On January 6, 2004, the comment period on Griles' draft EIS for mountaintop-removal mining closed. The next day, before all the comments could possibly have been read, the administration published its proposal to change the stream buffer rule. On January 7 the Federal Register contained a notice proposing to abolish the rule. The proposal is likely to become law before the November election. Joan Mulhern of Earth Justice, an environmental group that helped generate tens of thousands of comments, said, "They had obviously already sent the rule change to the Federal Register before the comment period had even closed, because it takes a couple of days for proposed rules to get printed. Clearly the Bush administration doesn't care what the public thinks."

"When it comes to the coal industry," Cindy Rank of the West Virginia Highland Conservancy told me in frustration, "they don't even need to lobby anymore. With Griles in there, it just happens."

The pillage of Appalachia by the coal industry is being made possible by officials who view public service as an opportunity for wholesale plunder. It is just one tragic legacy of this White House. "I believe that the coal industry has found the best friend they've ever had in the Bush administration," Judy Bonds told me. "Definitely the Bush administration and the coal industry have teamed up to wipe Appalachia off the map. This is Appalachia's last stand. When the mountains go, so goes our culture and our people, and it'll be the Bush administration that pushes the stake through our heart."

8

Killing the Energy Bill

The handouts to King Coal were bad enough. But we knew things could still get worse, much worse: If Dick Cheney's energy bill came to pass, the giant subsidies and rollbacks doled out ad hoc to the coal industry would be multiplied several times over, with catastrophic impacts on America's economy, environment, and values.

The collective horror at Dick Cheney's plan galvanized the environmental movement. National groups stepped up their efforts to defend the United States against Bush's overall agenda, while a core faction formed the Energy Strategy Group, specifically to derail Cheney's energy package. I worked alongside the leaders of this team, NRDC legislative director Karen Wayland and her predecessor, Alyssondra Campaigne. The group included energy specialists from the NRDC, the Sierra Club, and the U.S. Public Interest Research Group. We ferreted out donors, conducted polling, mobilized members to contact their public officials, and did our best to alert the public to Cheney's plan to loot the country's treasures. We deployed the NRDC's influential "network of aces"—business leaders, actors, and former political leaders who could call senators directly. More than 500,000 activists wrote letters and sent e-mails to their elected representatives and the White House. The NRDC sent me on a media blitz, writing editorials and articles, doing back-to-back interviews with television, radio, and news outlets across the country. The spirit of cooperation made this one of the best collaborative efforts in the history of the environmental movement.

Some groups, like Friends of the Earth and U.S. PIRG, concentrated on following the money, tracing the path from corporate contributions to lucrative legislation. The Sierra Club tapped its broad national membership to alert editorial boards and members of Congress at district meetings. The Wilderness Society and the National Environmental Trust handled the analysis of the bill's catastrophic impact on public lands. The NRDC, the Sierra Club, and the U.S. PIRG deployed their experts to scrutinize every other aspect: efficiency and conservation, coastal issues, coal-bed methane, clean air policy, Indian reservation energy issues, drinking water, global warming. And for 18 months, while Democrats controlled the Senate, we succeeded in repelling the Cheney attack. Senate Majority Leader Tom Daschle wouldn't let the bill move.

Then, in January 2003, Republicans took control of both houses of Congress. That April, House Republicans, led by Majority Leader Tom DeLay and Energy Committee Chair Billy Tauzin, passed a nightmare version of the Cheney package.[1] The Senate's new Republican leadership, however, decided that their companion bill, hobbled by polarizing elements of the Cheney plan, could not pass a floor vote. That's when the Democrats got snookered.

Setting up a bait-and-switch, the Republicans floated the idea that they might reconsider a Democratic energy bill that had stalled just after the 2002 election, while the Democrats still controlled the Senate. It was a bill that the environmental community didn't much like, but it wasn't the house of horrors that the current one was. The Democrats entered the trap. Democrat Harry Reid of Nevada, who had the floor, offered the previous year's bill. The Republicans rallied behind it as promised, and it passed on July 31.[2]

A chagrined Karen Wayland was watching on C-SPAN from her office. "At the moment the bill passed, Pete Domenici, the New Mexico Republican who chairs the Senate Energy and Natural Resources Committee, grinned widely and said, 'I'm happy because I'll be rewriting that bill. It's up to us, we're in the majority, and we'll be writing a completely new bill.'"[3] Domenici proceeded to take the bill to "conference." Whenever companion bills get approved by the two branches of Congress, the leaders of both houses work together in conference to reconcile the two into a final version that both branches must pass with a majority vote. But Domenici knew that in this conference committee, the Republicans, who controlled both houses, would be all alone. He could rewrite the Democratic bill from scratch. The Democrats had sealed their fate by failing to insist on a provision preventing the Senate bill from being changed in conference.

Over the next several months, the Republicans took maximum advantage of the situation, excluding Democrats while they constructed an energy-lobby dream deal. NRDC political coordinator Greg Wetstone, who had previously been counsel for the House Committee on Energy and Commerce, watched the spectacle from the sidelines. "They didn't let one Democrat participate in the process," he remembers. "They abandoned all the old models that included working together with people, soliciting bipartisan support, forming coalitions, reaching out to the other side, and negotiating compromises where everybody's voice is heard. The only outsiders with input were energy-industry lobbyists. And at a time when we desperately need a national energy policy, they produced a special-interest bonanza."

Since the process was completely controlled by the Republicans, none of us saw the bill as it was being hammered out, but we heard rumors from some of the moderate Republicans. Among the Republicans' shrewdest maneuvers was to keep, until the bitter end, two particularly offensive items in the bill: a proposal to drill oil in the Arctic National Wildlife Refuge and another to lift a moratorium on oil exploration on the outer continental shelf. Recalls Wetstone, "They forced us to spend our chits with the moderate Republicans fighting those hot-button issues." In the end they could claim, as Billy Tauzin did at the press conference announcing the bill, that they'd dropped those provisions as a concession to environmentalists, fostering the illusion that the bill was a product of some process of negotiation. Even National Public Radio's Terry Gross was taken in. When I was a guest on her show *Fresh Air* during the debate, she chided me that the bill couldn't be so one-sided since the Republicans had dropped plans to drill in the refuge. "It was like putting lipstick on a donkey," I responded.

Senator Domenici and Representative Tauzin finally announced on Friday, November 21, 2003, that the bill was ready.[4] The House vote was scheduled for the following Tuesday and the Senate's was set for Wednesday. Domenici was coy when reporters asked him to release the bill at his Friday press conference. In a clear attempt to minimize media attention and public scrutiny, the Republican leadership refused to release the text of the bill until late Saturday night, giving congressional Democrats a mere three days to review the 1,200-page document before voting on it.[5]

The NRDC's rapid-response team spent Saturday night and Sunday picking the bill apart. Wayland assigned specific sections to each of the NRDC's experts, and, after retrieving their analysis, she consolidated the reports for distribution to other environmental groups, the press, and our allies in Congress.

The thing was hideous. There were more than 60 loathsome provisions that would damage the environment and add billions to the national debt. It would establish oil and gas development as the dominant use of federal lands, subsidize the building of more nuclear power plants, and exempt polluters from core provisions of America's clean air and water laws. The best estimate of the bill's corporate tax breaks and subsidies was $100 billion. Republican Senator John McCain called it the "no lobbyist left behind bill."[6] It had more pork than a Smithfield slaughterhouse. Even factory farm multinationals got a 50-cent-per-gallon tax break—for using diesel fuel. Naturally, Halliburton got its share of the booty—a tax break for its environmentally destructive hydraulic-fracturing technology. There was also money for a shopping center in Louisiana that would host a Hooters, leading McCain to dub the bill a welfare giveaway "for Hooters and polluters."[7]

"This was legislation by unrestrained greed," says Wayland. "Everybody got their piece."[8] Wetstone, the coordinator of the environmental community's fight against Newt Gingrich's notorious "Regulatory Reform" bill back in 1995, told me, "In 25 years, this is the worst bill for the environment that has ever proceeded through the process. It absolutely devastates America's public lands and our existing laws protecting air and drinking water. It increases our dependence on foreign oil, bankrupts our Treasury—and it's antithetical to making progress on the most important environmental problem we've ever faced, which is global warming." Speaking of the Republicans in conference, Wetstone shakes his head in disbelief. "It was like Humphrey Bogart in *The Treasure of the Sierra Madre.* They got gold fever. There was never enough. They loaded the pork train down with favors for every polluter. Then they piled on money for every congressman who would support the bill and tried to keep it moving fast enough to avoid scrutiny and debate."

Among the biggest subsidies were multibillion-dollar packages for ethanol manufacturers, included to lure support from farm-state Democrats.[9] That tactic prompted a series of demoralizing defections by Midwestern Democrats, including our former champion, Tom Daschle.[10] Daschle had long promoted ethanol as a salve for South Dakota's economic woes. He was in a very tight race, and his in-state political advisers told him that if he didn't support the energy bill with its ethanol provisions, he was politically dead. But at least Daschle refrained from taking a public position on the bill until the last day, leaving other Democrats with the flexibility to join us on the barricades without appearing to challenge his leadership.

Ethanol lost us other core Democrats: Kent Conrad of North Dakota, Tom Harkin of Iowa, Tim Johnson of South Dakota, and Mark Dayton from Minnesota.[11] Two more perched on the fence until the last day—Evan Bayh of Indiana and Dick Durbin of Illinois. From the outset, we'd been hard-pressed to figure out how to block this legislation without the ability to control the gavel in the House or the Senate and without a veto pen. Now we were dealing with an entirely new political landscape, scrambling to cultivate new constituencies.

Our only weapon was the filibuster. Senator Charles Schumer of New York, along with Senators Dianne Feinstein and Barbara Boxer of California and Senator Mary Cantwell of Washington, had promised that they would use Senate rules to mount an endless debate to block the bill.[12] The Republicans would need 60 votes to end debate. But Cheney was lobbying furiously, calling senators, trying to cut into our 41 reliable filibuster votes. Senate offices were flooded with calls from energy and agricultural lobbyists and others with an interest in the bill.[13]

The White House and Republican leaders were using the entire omnibus budget bill to buy off votes. "They had the whole federal budget to bribe supporters," recalls Wetstone. Domenici, Senate Majority Leader Bill Frist, and the other senior leadership were coaxing Democrats onto the pork train, offering lucrative projects in their home districts. It got so bad that the *Wall Street Journal* condemned GOP senators for engaging in "months of plotting to buy enough votes with some $95 billion in tax breaks and pork-barrel spending."[14]

When the price was right, even some of our stalwarts went south. Byron Dorgan of North Dakota had held a press conference two weeks before the energy bill was released at which he vowed passionately that he would stand firm for a true national energy policy that must include vital programs for conservation; including, for example, the requirement that utilities have renewable electricity in their portfolios. Dorgan is from a corn state but swore he wouldn't be bought off by ethanol. "I've been working on ethanol for a long time, and the ethanol provisions as they're being discussed right now don't meet my requirements for what an ethanol mandate should be, so that's not going to buy my vote." Afterward, the Republicans added an $800 million coal gasification power plant in his state and Dorgan switched his vote. So much for standing firm for a national energy policy.

NRDC members and other activists sent more than 100,000 messages to Congress, urging the bill's defeat. I hit every talk show that would have me and stayed on the phone to Democratic senators from Saturday through Thursday. Key Midwesterners like Bayh and Durbin told me they couldn't commit until they saw the final bill. There was a storm of e-mails back and forth. "It was a really wild, exciting ride," recalls Karen Wayland. "We felt like David going up against Goliath. I mean it was such a hostile political climate. It just seemed impossible that we could win."[15]

Evan Bayh told me on Tuesday that he would oppose the bill and weather the wrath of Indiana's powerful corn lobby. When I thanked him for his courage, he dismissed the compliment. "It was the Hooters," he said. Indiana is in the Bible Belt, but Bayh joked that it wasn't a moral objection: "They were pissed that Indiana didn't get a Hooters."[16]

The bill was like old fish: It didn't take long before it began to stink. "It was classic hubris. They got greedy, they couldn't stop themselves, and they overreached," said Greg Wetstone. "Once it got examined, we got six Republicans to stand with us."[17] We had always hoped for support from the New England moderates Olympia Snowe and Susan Collins from Maine, and Lincoln Chafee from Rhode Island. But we also got three conservatives: Arizona's John McCain along with John Sununu and Judd Gregg from New Hampshire.[18] "I would say we all had concerns on both sides," Sununu later told me. "McCain, Gregg, and I were probably the most appalled by the fiscal issues and Collins, Chafee, and Snowe were most appalled by the environmental issues. But there's no question our interests overlapped significantly in both areas.

The bill seemed to be designed to try to address the needs of specific elements of the energy industry rather than get out with good ideas for producing reliable energy."[19]

On Tuesday the House passed the Domenici bill.[20] Then they tried to ram it through the Senate. But the vote never came. The Republicans failed to muster the 60 votes needed to kill our filibuster. As dozens of the bill's most destructive provisions came to light—many of them exposed by the NRDC's legislative team—a national tidal wave of editorial opposition and public outrage swamped Congress. Even the *Wall Street Journal* joined the din against the bill, calling it "a cornucopia of special interest energy payoffs."[21] Unable to secure the votes they needed to end the filibuster on Wednesday, the Republican leadership rescheduled the filibuster vote for the following morning.

A group of my colleagues—NRDC president John Adams, Greg Wetstone, energy guru Dan Lashof, and Karen Wayland—were working the phones and the Senate lobby when it came time to vote on Thursday morning. Everyone expected the Republicans to succeed in killing our filibuster. The energy industry was so sure of victory that the American Petroleum Institute had planned a huge celebration at Charlie Palmer's, Washington's cigar-chomping, good ol' boys' steakhouse on Friday night after the vote. There were a few environmentalists and a multitude of industry lobbyists milling around the anteroom and the Capitol's adjacent hallways. "We went into that vote still not knowing where three or four members were," Wetstone told me. Then the vote to end the debate was taken.[22]

We won by two votes. Despite last-minute arm twisting, with our six Republicans the Senate leadership fell short of forcing an end to debate on the bill.[23] "It was a very special moment, a huge victory for us against enormous odds," Karen Wayland says of that night.

I was at the National Press Club in Washington, D.C., addressing a banquet when Wetstone called me with the news. When I announced it a few minutes later from the stage, there was a wild ovation.

Wetstone remembers the glum industry lobbyists who were milling about the Senate lobby looking depressed. "They were shocked," he says. "They had to cancel their victory party at Charlie Palmer's." He recalls watching groups of them talk heatedly after the vote: "Some of them had made hundreds of thousands of dollars telling people they're going to bring home these massive taxpayer subsidies and special gifts for polluters." Wetstone sounded almost like he felt sorry for them. I reminded him that they still got paid.[24]

I ran into Senator Sununu that evening. He was leaving the CNN building in Washington and hailed me as I walked in to record *Crossfire.* He was bouncing with energy, stoked by the victory, and he promised me that even if Domenici broke the filibuster and brought the bill back the following week—which environmentalists expected and feared—he would block it on the budget point of order, a procedural objection that requires a separate vote on fiscal issues. Any member can call for such a vote when it appears that a bill would break the statutory budget cap. Normally, if there's support for a bill, the Senate just ignores that rule. But this time, Sununu would raise the objection. He also assured me that we would pick up votes from many conservative senators who had voted for the bill to support the president but whose principles would compel them to oppose the bill once it was formally framed as a budget issue.

"As fiscal conservatives, we have to stand for something,"

said Sununu. "This bill broke the budget caps for 2004 and busted budget caps over a five-year period. And the tax provisions were outrageous. The agreement had over $24 billion in tax breaks. There's just no need to provide special consideration to the oil, gas, or coal industries or the nuclear power industry," he continued. "These are healthy, robust, competitive industries, and they don't need special treatment or a special tax break from the federal government. Certainly not beyond those that are already in place. Plus, the MTBE provision was outrageous!"

Thank God for the MTBE waiver. That provision would have immunized the oil industry from liability for contaminating water supplies in Sununu's home state of New Hampshire and across the country with methyl tertiary-butyl ether, a gasoline additive that was introduced in 1978 to prevent "knocking."[25] It was tacked onto the pork train by the right-wing oil-patch cabal of Tom DeLay, Billy Tauzin, and Texas Congressman Joe Barton, chairman of the Energy and Air Quality Subcommittee.[26] The MTBE provision is a microcosm of all that was wrong with Dick Cheney's energy bill, an emblem of greed that finally broke the bill's momentum.

I discovered the horrors of MTBE in 1999, when, with help and advice from actor Paul Newman, I launched a bottled-water company, Keeper Springs, which donates all profits to environmental activism. Keeper Springs water comes from a pristine spring in the Green Mountains of Vermont. I quickly learned, however, that every springwater company lives in perpetual fear of finding MTBE in its aquifers. MTBE, which leaks from underground gasoline storage tanks, causes kidney and liver cancer and possibly leukemia, lymphoma, and testicular tumors in laboratory animals.[27] Even the minutest quantity renders water so foul-tasting that people can't drink it.

In 1978, ARCO Chemical submitted a request to the EPA for a waiver to use MTBE, an oxygenate, to replace lead in high-octane gasoline. MTBE is a by-product of the refining process, and the oil companies were eager to squeeze some profit from it. It wasn't long before it was being added to nearly all gasoline. From the beginning, the industry knew there were problems. As early as 1984, an internal Exxon memo acknowledged, "we have ethical and environmental concerns" about using MTBE.[28] Industry surveys as early as the 1980s suggested that between 15 and 40 percent of underground gasoline storage tanks at service stations were leaking.

In June 1984, the American Petroleum Institute created the MTBE Task Force, which included representatives from Shell, ARCO, and Texaco, to strategize about the huge concentrations of the chemical being found in groundwater. It was often discovered in wells where there was no detectable gasoline, meaning that it was moving farther and faster through soils than other compounds. While other gasoline components bind to soils and tend to remain close to a spill, MTBE is soluble and moves with the water for greater distances. MTBE, the task force acknowledged, cannot be removed by conventional treatments like carbon absorption, making it extraordinarily difficult to remediate.

In 1990, MTBE could be found in groundwater across the nation. In October of that year, Shell called its industry sisters to an MTBE environment meeting in The Hague to warn that MTBE "is not biodegradable in water" and that the oil industry should anticipate expensive regulations to redress damages.[29] Shell executives joked in an internal document obtained in one of our recent court cases that MTBE stood for "Most Things Biodegrade Easier," or "Menace Threatening our Bountiful Environment," and, finally, "Major Threat to Better Earnings."[30]

Yet the oil companies actively concealed this information from state and federal regulators as they continued to promote MTBE's use nationally. In 1991 the industry successfully pressured the EPA to allow the companies to use MTBE to meet new emissions standards in the Clean Air Act—despite the availability of safe, cheap alternatives like ethanol.

A nationwide study conducted by the U.S. Geological Survey between 1993 and 1994 found MTBE in 27 percent of urban wells.[31] By 1997, large MTBE concentrations had forced the closure of public and private wells in dozens of cities and towns across the United States. The costs of treating MTBE contamination in the country's drinking water have been calculated to be $29 billion, with some researchers estimating as much as $46 billion—a bill that the oil industry is trying to dodge.[32]

On the theory that it's always cheaper to fix the law than the problem, the industry mobilized its powerful lobbyists to push the White House and Congress for protection.[33] Their request ended up in the energy bill.

The MTBE provision may rear its ugly head again, however. The big energy companies, as they watched their dream bill go down in defeat, had too much to lose to simply accept defeat on MTBE—or on anything else. "This fight isn't over," said John Adams, NRDC founder and president. "There are so many billions of dollars at stake that it's kind of hard to believe they're not going to find a way to make this happen."

9

National Security

On March 2, 2004, President Bush gave a speech to mark the first anniversary of the creation of the Department of Homeland Security. With a gunslinger's bravado, he noted that the crack experts in his cabinet had taken "unprecedented measures to protect the American people here at home," noting that his staff deserves "a gold star for a job well done."[1]

Just across the Hudson from New York City, a few miles from Ground Zero, is the Kuehne Chemical Company of Kearney, New Jersey, one of the country's largest producers of chlorine products.[2] In 1999, Kuehne filed a risk-management plan with the EPA. In accordance with the Clean Air Act, every company that uses or stores extremely hazardous chemicals is required to file these safety reports at least every five years. The Kuehne report offered a chilling assessment of a worst-case scenario at one of these plants: "Fully loaded railroad tank car releases all its chlorine within ten minutes. The resulting cloud of chlorine vapor would be immediately dangerous to both life and health for a distance exceeding fourteen miles. The total population in this radius is approximately twelve million."[3] There are 15,000 such facilities in the United States, including an estimated 111 that, if attacked, could each put a million or more people at risk of death or injury.[4] Eight of these are in New Jersey.

And what "unprecedented measures" has Bush enacted to prevent this horror from occurring? Next to none. Before September 11, there was almost no government oversight of the security precautions taken by these industries—and there still isn't today. Since September 11, the White House has done nothing to require better security at those 15,000 chemical manufacturing facilities, oil tank farms, pesticide plants, and other repositories of deadly chemicals. Nor has it forced the nuclear industry to beef up security adequately at its 103 nuclear power plants. Three years after employees watched plumes of smoke pouring from the Trade Towers just across the river, there has been little verifiable change in the safety measures at the Kuehne Chemical Plant.[5]

While the Bush White House is engaged in a foreign policy seemingly designed to recruit terrorists and encourage more attacks against our nation, it is doing little to require corporations to protect Americans from those attacks.

The risks associated with the nation's many chemical facilities are disputed by no one. "The impact of a terrorist attack on a chemical facility could overshadow the human and economic costs of the World Trade Center attack," says Rand Beers, President Bush's former director for combating terrorism and a White House counterterrorism adviser for 30 years.[6] Troy Morgan, the FBI's specialist in weapons of mass destruction, calls chemical tank farms the "poor man's atomic bomb."[7] The Homeland Security Department, the Justice Department, the GAO, and the U.S. Army surgeon general, as well as a host of independent organizations, all conclude that chemical plants are attractive terrorist targets and potential weapons of mass destruction. In February 2003, the National Infrastructure Protection Center, now part of the Department of Homeland Security, warned that Al Qaeda or other terrorist groups might "launch conventional attacks against the U.S. nuclear/chemical-industrial infrastructures to cause contamination, disruption and terror."[8]

Seven weeks after the September 11 attacks, Senator Jon Corzine of New Jersey introduced the Chemical Security Act, a bill that would require chemical plants to reduce toxic chemical inventories where practical and switch to less toxic chemicals when it was economically feasible.[9] In many cases, Corzine told me, a plant manager could store a 30-day inventory of toxic chemicals rather than a full six months' worth. "Simple, sensible economic reforms of that kind could save a lot of lives in a catastrophic event."[10] The bill passed the Environment and Public Works Committee in a 19–0 vote.

But then the chemical lobbyists went to work. During the following August recess, before the bill went up for a Senate vote, the chemical-industry trade associations began flooding senators' offices with requests to kill the Corzine bill. Leading the fight was the American Chemistry Council. The ACC's counsel, James Conrad, served on President Bush's EPA transition team. Fred Webber, ACC's former president, is a Bush Pioneer and old friend from Texas who helped recruit 25 other chemical industry executives to be Bush fundraisers. According to Common Cause, the ACC and its member companies contributed more than $38 million to Republicans between 1995 and June 2002 and spent another $30.2 million on lobbying during the same period. In July, August, and September

2002, while they were actively fighting the Corzine bill, members gave an additional $1.3 million in PAC contributions to soften up our public servants.[11]

The chemical industry's furious lobbying included ads in Capitol Hill newspapers and op-eds branding the Corzine bill as a subversive plot against American industry. The National Propane Gas Association called it "Stalinesque."[12] Conservative think tanks marched in lockstep with the trade associations; the Heritage Foundation said that the supporters of Corzine's bill were not really intent on fighting terror, but "have a different agenda: taking a large step toward their goal of a chemical-free world."[13] Angela Logomasini of the Competitive Enterprise Institute said that Corzine's bill was "designed to serve a radical environmental agenda that targets chemicals."[14] Amy Ridenour of the National Center for Public Policy Research said that Corzine and his radical environmental friends have mounted a "jihad against the chemical companies."[15] Then there were the predictable knee-jerk ideological denunciations of the EPA by the Wise Use crowd: "Our experience with the EPA [is]," said Rebeckah Freeman of the American Farm Bureau, "if you give them an inch, they take 10,000 miles."[16]

A baffled Corzine argued that his bill wasn't even about environmental protection: "This is a safety and public protection issue that is no different than making sure people don't take guns on airplanes."

Eight Republican senators, including Senator James Inhofe of Oklahoma, who became chairman of the Environment and Public Works Committee when his party took control of the Senate, announced objections to the measure in letters to their colleagues.[17] Their rationales parroted the chemical industry's arguments that Corzine's bill "may not adequately build on" the chemical industry's existing "initiatives . . . that form a sound foundation to improve security." These eight senators, by the way, had received more than $750,000 from ACC members and their PACs since 1995.[18] By the autumn, Corzine's bill was dead.

Meanwhile, EPA Administrator Whitman was attempting to respond to the public outcry over lax security at chemical plants. The Clean Air Act empowers the EPA to force companies to safeguard the public from accidental chemical releases. But the agency had never issued regulations requiring precautions against a terrorist attack. After September 11, while Corzine was trying to push his bill through the Senate, the EPA staff developed a plan to use the agency's authority to require improvements in chemical plant security. The suggested reforms echoed Corzine's; facilities ought to reduce toxic inventories and use less dangerous chemicals where feasible. If plants can convert to safer chemicals or processes that cannot be used as weapons of mass destruction, it obviously lessens the daunting task of guarding those facilities. The late Jim Makris, then-chief of the EPA's Chemical Emergency Preparedness and Prevention Office, said the goal would be to make it habitual among company management to think about ways to make their plants safer.[19]

The chemical lobby also opposed the notion of requiring plants to use safer technologies. Greg Lebedev, then-president of the ACC, took a stab at explaining industry fears about regulations in an interview with New Jersey's *Bergen Record.* "Chemical companies make dangerous things. Getting into the technology of what you make and how you make it is a subject for an environmental or technology context, not security. I don't want to wander down an exotic path here."[20]

Whitman was caught in the squeeze between the public clamor for action and her reluctance to confront the chemical industry and powerful White House friends. So instead of using its existing authority, the EPA worked with the White House for a year trying to hammer out a regulatory proposal that wouldn't anger industry. That effort finally died due to White House intransigence.

Still, Whitman was under pressure to do *something*—there was too much noise in the press about the threat from chemical plants. So she decided to beef up inspections. The EPA has the authority to inspect chemical plants at will, without permission or advance notice. But Whitman wasn't prepared to antagonize the White House. So with the confidence that marked her tenure at the EPA, she boldly asked permission of the nation's 30 highest-risk chemical plants to allow inspectors to visit their facilities.[21] Some companies refused outright, and the chemical moguls called in their White House chits to end the EPA's meddling once and for all. In early 2003, under industry pressure, the White House yanked the EPA's authority over chemical security and transferred oversight to the Department of Homeland Security.[22] From the industry's perspective, the DHS was the perfect overseer. A fledgling department with no expertise to inspect chemical plants or legal authority to require the chemical industry to implement tougher security standards, the DHS was unlikely to impose new burdens. Since then, the federal government has not troubled the industry about security.

"There are currently no federal security standards for chemical facilities," Corzine told me. "None at all. The industry does what it desires or what it thinks it can afford—and millions of Americans are at risk."[23]

The argument advanced by the chemical companies after September 11—and the one that the Bush administration apparently swallowed—is that, left alone, producers will take voluntary measures sufficient to protect the public. The ACC claimed that its own chemical security code, which is mandatory for its members, would lead to tighter chemical security than would Corzine's bill. Yet only 7 percent of the 15,000 chemical facilities in the United States are members of the ACC.[24] Moreover, the voluntary plans that do exist are anything but stringent. Sal DePasquale, a former Georgia Pacific official who helped write the ACC's security plan, wrote in a letter to the U.S. PIRG that voluntary standards are "a smoke and mirrors exercise to make it appear that it is issuing bona fide standards. . . . Across the country there are huge storage tanks with highly dangerous materials that are far from adequately secured."[25] Even Homeland Security Secretary Tom Ridge has stated that "voluntary efforts alone are not sufficient."[26]

You can say that again. In April 2002, a *Pittsburgh Tribune Review* reporter named Carl Prine wrote of how he was able to enter 60 dangerous chemical plants virtually unchallenged.[27] In Baltimore, Houston, and Chicago, he strolled through unguarded gates in broad daylight, wearing a press pass and carrying a camera. He drove up to tanks, pipes, and control rooms considered key terrorist targets. Hardly anyone tried to stop him. He found security nonexistent in many places. "I walked into one Chicago plant," he told me. "I climbed on top of the tank and sat there and waved, 'Hello! I'm on your tank.' I wondered what it would take for me to get arrested at one of these plants. Would I have to come in carrying an AK-47? What would it take for someone to say 'Why is this guy walking around taking pictures of our tanks?'" Reporters from all over the country took Prine's lead and filed stories about their infiltration of dangerous plants.[28] Secretary Ridge acknowledged these reports when he testified before the Senate Environment and Public Works Committee on July 10, 2003, noting that there were deficiencies at "dozens and dozens" of U.S. chemical facilities. "Our enemies," he warned, "look at [chemical plants] as targets."[29]

To the right-wing radicals in the White House, laissez-faire capitalism is the legitimate response to every contingency—even national security. On March 21, 2003, PBS reporter Daniel Zwerdling questioned Tom Ridge's top aide, Al Martinez-Fonts, about Homeland Security's reluctance to mandate security reforms in the chemical industry beyond voluntary programs.[30] Martinez-Fonts, a former executive of JP Morgan Chase, said that, even in a time of war, the Bush administration was reluctant to interfere with business decisions by the private sector.

"I was in the private sector all my life," explained Martinez-Fonts. "Did I like it when the government came in and stepped in and told [us] to do certain things? The answer's no. In general, we don't like to be told what to do. . . . The administration has been very proactive towards business, promoting business issues, et cetera. The point is, people are concerned that a lot of regulation, a lot of legislation might ultimately come out. I think we're trying to avoid that. I, as the person representing the private sector in Homeland Security, would prefer to avoid that altogether."

When Zwerdling reminded Martinez-Fonts that the federal government told the airline industry to improve its security and asked whether it doesn't make sense for the government to also require security upgrades by the chemical industry, Martinez-Fonts replied: "Well, the answer is because September 11 happened, and they were airplanes that rammed into buildings. And it was not chemical plants that were blown up."

So much for homeland security!

* * *

Of course, toxic chemical plants aren't the only potential dirty bombs on American soil. The nation's nuclear power plants pose an equally devastating threat. The most vulnerable one happens to be in my backyard. I live in Mount Kisco, New York, 11 miles downwind of the Indian Point Nuclear Power Plant. Indian Point's two remaining active reactors—Unit 1 was shut down in 1974—sit on the east bank of the Hudson River, 24 miles north of New York City.

On the morning of September 11, 2001, United Airlines Flight 175 from Boston passed within a few thousand feet of Indian Point as it followed the Hudson River down to its rendezvous with Tower Two of the World Trade Center. Had it banked left and crashed into the plant instead, it could have triggered a large release of radiation. The surrounding area, including New York City, might have been rendered uninhabitable for years.

Neither the NRDC nor the Hudson Riverkeeper have ever taken a stand against nuclear power, but following the terror attacks, the communities surrounding Indian Point inundated our offices with phone calls and letters expressing concern about plant safety. The plant, however, carried on, business as usual.

Meanwhile, a few miles from my home, busy roads were closed to prevent anyone from getting near upstate reservoirs. Sport fishermen—a major economic resource to the region—were ordered off the reservoirs, and subsequently the bait-and-tackle shops shut down. The proprietor of Bob's Tackle Shop in Katonah was stoic about closing her family business, but wondered why Indian Point was still chugging along. "It's crazy," says Captain Ron Gatto, the top cop in the city's upstate reservoir. "There is no way a fisherman could sabotage the city's water supply—you'd need tanker trucks filled with poison.

Everyone knows that the biggest threat is Indian Point. I lose sleep knowing how vulnerable this whole system is. It's absolutely insane—they're only open 'cause they've got pals in Washington."[31]

This outcry prompted us to study Indian Point in light of the risks of terrorist attack. No nuclear facility in the United States is closer to such a densely populated metropolitan area.[32] Captain Gatto is not the first person to use the word "insane" to describe Indian Point. Commenting on the siting of Indian Point in 1979, in the wake of the Three Mile Island meltdown, Robert Ryan, the Nuclear Regulatory Commission's director of the Office of State Programs, stated: "I think it is insane to have a three-unit reactor on the Hudson River in Westchester County, forty miles from Times Square, twenty miles from the Bronx. And if you describe that fifty-mile circle, you've got twenty-one million people. And that's crazy. I'm sorry. I just don't think that that's the right place to put a nuclear facility."[33]

Contrary to the public perception aggressively promoted by the industry, terrorists would not have to puncture the containment dome to cause a serious accident or meltdown. Nuclear plants like Indian Point are vulnerable at half a dozen points, some of them virtually impossible to shield from determined attackers. Terrorists could provoke a meltdown by coordinating attacks against the reactor's cooling system or the plant's control room, by cutting electric lines going into or out of the plant, or, more alarmingly, by disabling the cooling-water pumps and intake structures, which are easily approached from the river's channel.

Worst of all, in a catastrophe that would rival or exceed the impact of a meltdown, terrorists could attack the plant's spent fuel pools, which house 30 years of accumulated high-level radioactive waste and are shielded only by a series of flimsy annex buildings, so-called butler shacks that have the structural integrity of a Kmart. Indian Point's irradiated spent-fuel pools contain more than 1,500 tons of high-level radioactive waste.[34] According to the U.S. Nuclear Regulatory Commission (NRC), a significant loss of water within the spent-fuel pools could provoke a fuel-assembly fire that could potentially release a pool's store of cesium 137—up to 20 times the amount released at Chernobyl, which made an area approximately 1,000 miles around the plant uninhabitable, 100 miles of it permanently.[35]

Imagine a world without New York City. The terrorists have. The Al Qaeda network and other groups have cited nuclear power plants as potential U.S. targets. President Bush warned us during his 2002 State of the Union Address that Al Qaeda terrorists possess diagrams of U.S. nuclear facilities. On the CBS news program 60 *Minutes II,* Yosri Fouda, a reporter for the Arabic news network Al Jazeera, said that when he interviewed the recently captured Khalid Shaikh Mohammed, architect of the September 11 attacks, in the spring of 2002, Mohammed said that nuclear facilities in the United States were Al Qaeda's first choice of a target.[36] In November 2002, the FBI warned that Al Qaeda sleeper cells could be planning attacks on U.S. nuclear power plants near our largest cities to try to inflict "severe economic damage and maximum psychological trauma."[37] Indian Point's proximity to the world's financial center, and the severe consequences for public health, national security, the environment, and the economy in the event of a successful terrorist attack make that plant especially attractive.

With this in mind, you would think, and most people do, that in a country as civilized and technologically advanced as ours, nuclear plants would be among our most secure facilities. Amazingly, the opposite is true. Indian Point and other plants near Chicago, Cleveland, Detroit, Miami, Minneapolis, Philadelphia, Pittsburgh, Phoenix, and Washington, D.C., are virtually unprotected against terrorist attack on the scale of September 11.[38] If you think this sounds like an exaggeration, consider this astounding fact: Federal law absolves nuclear power plant operators from any legal duty to protect their plants from attacks "by enemies of the United States."[39] So who does shoulder this heavy burden? Governor George Pataki of New York tells us that it is the federal government. But try to find a federal agency that will take responsibility. Not the NRC, not the Department of Homeland Security, and not the Pentagon.

Nuclear plants are required to show that they can resist attacks by small groups of vandals who are not "enemies of the United States." And the NRC periodically conducts mock attacks by such saboteurs, typically sending small forces of two or three attackers. Astoundingly, nearly 50 percent of the nation's nuclear facilities routinely fail to repel even these feeble assaults, despite being notified of the time and date of the attack months in advance.[40] Indian Point is apparently among the most poorly defended of the entire nuclear fleet. In a 2002 internal report by Entergy Nuclear, the plant's Mississippi-based owner, obtained by Riverkeeper, Indian Point's security guards acknowledged that the robust security force portrayed by Entergy in its advertising campaign and public pronouncements is a deception. Mock attackers were able to enter the plant practically at will. The head of one of the teams, Foster Zeh, attested that he could breach the perimeter fence and place dummy explosives around the spent-fuel pools in under 40 seconds. According to the internal report, the guards are undertrained, underequipped, overworked, demoralized, and out of shape. Only 19 percent believe they could defend the plant from attack. Most of them said they would flee if the plant were attacked.[41]

A recent Riverkeeper lawsuit against the NRC before the Second Circuit U.S. Court of Appeals exposed "gaps" in plant security for the first time.[42] There is, for example, no protection from air attacks at Indian Point. The FAA has refused to declare a no-fly zone over the plant, which lies in the approach path of Westchester County Airport. The FAA has given this protection to Disneyland, Disney World, and Crawford, Texas. They even provided it for my cousin Caroline Kennedy's wedding on Cape Cod! In 2002, my brother Douglas, a reporter for Fox News, chartered a small airplane at Westchester Airport and flew directly over the plant with a film crew and circled it for 20 minutes, waiting in vain for someone to signal him off. Dr. Gordon Thompson, a research scientist at the Institute for Resource and Security Studies in Cambridge, Massachusetts, and an expert on nuclear plant security, told me that a small private jet, chartered at Westchester County Airport and packed with explosives by a sophisticated but suicidal terrorist, could crash into the right building and precipitate a spent-fuel fire—releasing all the plant's stored radiation.

There is also no defense against line-of-sight missile attacks from the west side of the Hudson, and only weak defenses against a marine attack on the plant's cooling-water structures, and even those are intermittent; a patrol boat is docked alongside the plant, usually a small Boston whaler or buoy tender. Last summer we were cavorting back and forth in front of Indian Point in the Riverkeeper boat to test its defenses, when two guards finally approached us in a whaler. When we asked whether they were armed, they sheepishly told us no, and explained that they would need to radio back for directions if there were an attack. Their boat broke down on its way back to the plant and the crew had to radio for help. Buoys mark an exclusion zone that recreational boaters generally respect. Terrorists could penetrate the zone and reach the plant in a matter of seconds.

But it's not just Indian Point that this White House is ignoring. On September 24, 2003, the GAO issued a report faulting the Bush administration for failing to bolster nuclear plant security nationwide.[43] The GAO found that the dereliction at Indian Point is in fact the rule at nuclear plants across the United States. According to the report, the NRC deliberately stages softball mock attacks to give the impression of plant security, and has often shielded the industry by burying significant security breaches.[44] NRC inspection reports routinely omit security violations such as a guard sleeping on duty or falsified security logs.

As it turns out, the sleeping-guard incident took place at Indian Point. The NRC report indicated that when two of its officials found a security guard napping at his post at the Indian Point 2 reactor last year, the agency decided not to issue a notice of violation because there was no terrorist attack on the plant during the half hour or so that the guard was sleeping.[45] The GAO auditors said that, nationwide, the NRC habitually refused to issue formal citations and routinely minimized the significance of problems it found if the problems did not cause actual damage (a circumstance that would occur only if terrorists happened to strike the plant when the NRC investigators were present). The NRC further explained that it elected to treat the Indian Point incident as a "non-cited violation" because no single guard had been found sleeping "more than twice during the past year."[46] Who says the NRC doesn't have a sense of humor?

Indian Point's license requires its operator to demonstrate that there is a workable evacuation plan in the event of an emergency. However, Entergy Nuclear is not required to develop a plan for the 50-mile peak injury zone—which would involve the impossible task of evacuating New York City. The company has, however, developed an emergency plan to evacuate the plant's 10-mile radius. It involves moving residents within the 10-mile radius to reception centers 11 to 15 miles from the plant. My local high school is a reception center. I doubt the evacuees will feel particularly safe there. Most of my neighbors intend to head for the hills as soon as the Indian Point emergency sirens sound. Entergy's evacuation plan is so comically absurd that my neighbor Chevy Chase seriously considered a stand-up routine consisting of reading excerpts from the document.

In August 2002, Governor George Pataki commissioned a consulting firm headed by former Federal Emergency Management Agency (FEMA) director James Lee Witt, the world's leading expert on emergency planning, to assess Indian Point's Emergency Response Plan. Witt's exhaustive 550-page report criticized virtually every aspect of the plan and concluded that Entergy's emergency plan is "not adequate to . . . protect the people from an unacceptable dose of radiation in the event of a release from Indian Point." Witt added that the current evacuation plans "do not consider the reality and impacts of spontaneous evacuation." His report pointed out that all emergency planning assumed a slow accidental release that could be kept secret for many hours in order to keep the roads clear while schoolchildren were evacuated.[47] Entergy apparently has not heard of cell phones or CNN.

Fallout from the Witt report was dramatic. Westchester's Republican congresswoman, Sue Kelly, who had been one of the few Congress members still defending the facility and its emergency plan, immediately joined approximately 260 elected officials, 35 municipalities, 56 environmental and civic groups, hundreds of business leaders, and several labor unions and school boards in calling for the shutdown of Indian Point. Four county governments (Orange, Rockland, Westchester, and Putnam) and the State of New York have refused to certify their evacuation plans to FEMA and the NRC. Adequate evacuation plans are a condition of a plant license but, not surprisingly, the NRC indicated it was willing to overlook an unworkable and unfixable emergency plan.

Despite numerous requests by Riverkeeper and local politicians, Homeland Security Secretary Tom Ridge has refused to meet with local leaders or take any position on the issue or investigate the matter. But Ridge went the extra mile to protect Entergy's profit margins. On July 25, 2003, FEMA and the Department of Homeland Security issued a determination that the emergency evacuation plan for Indian Point "would be adequate in protecting public health and safety in the event of a release."

The administration has independent power to shut down the plant under the EPA's Clean Water Act authority, and to require security and emergency preparedness improvements through the DHS, the NRC, the DoE, and FEMA. Entergy rakes in approximately $1 million a day from the electricity produced by Indian Point and has shared its profits generously with the president and his party. Entergy is a major player within the Nuclear Energy Institute (NEI). Entergy president Donald Hintz is chairman of the NEI's board of directors.

Contributions from companies and organizations on NEI's 2001 members roster total $29.2 million in soft money from 1991 to June 30, 2001, with 63 percent going to Republicans.[48] NEI itself contributed $643,202 during the same period.[49] It also spent nearly $10.8 million lobbying Congress and the executive branch from January 1996 through June 30, 2001. These have been great investments; NEI met with Energy Department officials 19 times while the Cheney task force was at work.

In her April 8, 2004, testimony before the September 11 Commission, National Security Advisor Condoleezza Rice swore under oath to the nation that the administration was doing everything in its power to "harden terrorist targets" in the United States. But, as we have seen, this administration is doing next to nothing.

The idea that industry will step up to the plate on its own is pure folly. In July 2003, the Conference Board, a business research group, found that American corporations have hiked security expenditures less than 4 percent on average since the September 11 attacks.[50] As terrorism expert Rand Beers told me, "Of course, there's no such thing as a perfect defense, but we're remiss if we don't try to protect our citizens. We know for sure, from interrogation of terrorists, that security measures have value, they discourage attacks. So, the idea that we should leave the barn door open, we should simply say *que sera sera*—the American people don't ask presidents or administrations to behave in that fashion. That is such a dereliction of duty as far as I could see." Beers' dissatisfaction with President Bush's inertia when it comes to protecting Americans prompted him to resign in March 2003, after providing counterterrorism advice to every president since Ronald Reagan.

Tragic as it was, September 11 was a gift of sorts to George Bush. As of August 2001, jobs were vanishing, the stock market was sinking, and the public was disgusted with Bush's policies on the environment. But after the terrorist attacks,

Bush's poll numbers soared. The public rallied to support its commander in chief as he sought to avenge the innocent lives we lost. But while the war against terror has required great sacrifices from American taxpayers—and vastly greater sacrifices from our fighting soldiers—President Bush has been unwilling to ask his friends and corporate paymasters to ante up their share. He has not challenged Detroit to end America's reliance on Middle East oil. He has not asked the Saudi princes to cooperate with the FBI. He won't ask the richest 1 percent to give up their tax cuts to help fund the war. He has not even attended a single funeral for our war dead. Instead he has used September 11 as a rationale to justify everything from the daily body-bag count to the swollen deficit. He has used the war on terror to implement Dick Cheney's neo-con agenda—invade Iraq, sink the UN, enrich Halliburton, reward the rich, open our lands to oil and mineral development, destroy our wetlands, hand obscene tax breaks and subsidies to the energy barons, and take a blowtorch to our constitutional rights and environmental laws—all in the name of homeland security.

Meanwhile, as our nation spends hundreds of billions of dollars searching for weapons of mass destruction in Iraq, we are ignoring 15,000 WMDs on American soil.

10

What Liberal Media?

For the last couple of years I've traveled around the country on an informal speaking tour, sounding the alarm about George W. Bush's record on the environment. I've spoken to hundreds of audiences, including conservative women's groups; public school teachers; civic, religious, and business groups; trade associations; farm organizations; rural coalitions; and colleges. As I talk about the plundering of our shared heritage, I urge these Americans to help protect the air and water, landscapes and wildlife, that enrich our nation and inform our character and values.

The universally positive response to my speeches confirms national polls that consistently show strong support for environmental protection across party lines.

But I invariably hear the same refrain from audiences: "Why haven't I heard any of this before? Why aren't the environmentalists getting the word out?" The fact is, there is no lack of effort on our part to inform the public, but we often hit a stone wall: the media. They are simply unwilling to cover environmental issues.

To some extent this has always been true. In 1963, President Kennedy and Senator Gaylord Nelson made a cross-country tour to alert Americans to the environmental crisis.[1] In speech after speech Kennedy warned that air and water pollution, species extinction, and pesticide poisoning were threats to our nation's future. But as he later complained to Nelson, the press asked only about national defense or power politics and never mentioned the environment in its stories. In fact, it was Nelson's experience on that trip that inspired him to organize the first Earth Day eight years later.[2]

Now the crisis that President Kennedy predicted is upon us. Ocean fisheries have dropped to 10 percent of their 1950s levels,[3] the earth is warming, the ice caps and glaciers are melting, and sea levels are rising.[4] Asthma rates in this country are doubling every five years.[5] Industrial polluters have made most of the country's fish too poisonous to eat. The world is now experiencing extinctions of species at a rate that rivals the disappearance of the dinosaurs.[6] Nearly 3 billion people lack sufficient fresh water for basic needs, and over 1 billion are threatened with starvation from desertification. Hundreds of millions of desperate people have been displaced by environmental disasters; the presence of these refugees puts added pressure on the local ecology, often leading to wars and further environmental degradation. All this at a time when our president is engaged in the radical destruction of 30 years of environmental law. These things are certainly newsworthy.

Yet it's hard to find much mention of this in the press. The Tyndall Report, which analyzes television content, surveyed environmental stories on TV news for 2002. Of the 15,000 minutes of network news that aired that year, only 4 percent was devoted to the environment, and many of those minutes were consumed by human-interest stories—whales trapped in sea ice or a tiger that escaped from the zoo.[7]

Why is the media barely covering such a vital public policy issue? Why isn't it informing the public and providing Americans the news they need in order to be effective citizens?

From the birth of the broadcasting industry, the airwaves—from which most Americans obtain their news—were regarded and regulated as a public trust, a communal resource like the air and water. The Federal Radio Act of 1927 required that broadcasters, as a condition of their licenses, operate in the "public interest" by covering important policy issues and providing equal time to both sides of public questions.[8] Those requirements evolved into the powerful Fairness Doctrine, which mandated that the broadcast media has a duty to maintain an informed public. Among other things, broadcasters had to air children's and community-based programming, and the rules were weighted to encourage diversity of ownership and local control. The Fairness Doctrine governed television and radio for most of the twentieth century.[9]

In the 1960s the Federal Communications Commission (FCC) and the courts applied the Fairness Doctrine to require cigarette manufacturers to include the surgeon general's warnings in their TV and radio advertisements, and polluters to notify the public when advertising a polluting product.[10] Advertisers of gas-guzzling automobiles, for example, had to provide rebuttal time for public interest advocates to debate the impact of wasteful fuel consumption on our environment and public health.[11] According to media commentator Bill Moyers, "The clear intent was to prevent a monopoly of commercial values from overwhelming democratic values—to assure that the official view of reality—corporate or government—was not the only view of reality that reached the people."[12] The U.S. Supreme Court unanimously upheld the Fairness Doctrine in the Red Lion case in 1969, confirming that it is "the right of the viewers and listeners, not the right of the broadcasters which is paramount."[13]

Then, in 1988, Ronald Reagan abolished the Fairness Doctrine as a favor to the big studio heads that had supported his election. The occasion was a case involving a Syracuse, New York, television station that had broadcast nine paid editorials advocating the construction of a nuclear power plant. When the station refused to air opposing viewpoints, an antinuke group complained. The three Reagan appointees who ran the FCC sided with the TV station, applying the same laissez-faire philosophy to the airwaves as the Reagan team did to the other parts of the common. They reasoned that the recent proliferation of cable TV allays the "Supreme Court's apparent concern that listeners and viewers have access to diverse sources of information." Broadcasters would henceforth be under no obligation to air views that opposed their own.[14]

Reagan's FCC chairman, Mark Fowler, scoffed at critics' concerns that the loss of the nation's most popular open forum diminished our democracy. "'Television," he said, "is just another appliance—it's a toaster with pictures."[15] A horrified Congress reacted with legislation codifying the Fairness Doctrine, but President Reagan vetoed the bills.[16] The FCC's pro-industry, anti-regulatory philosophy effectively ended the right of access to broadcast television by any but the moneyed interests.

As an unregulated part of the commons, TV and radio are today subject to the same dynamic that is polluting our other public trust assets, with behemoths consolidating control of and contaminating the airwaves.

One-sided and often dishonest broadcasting has replaced the evenhanded reporting mandated by the Fairness Doctrine. The right-wing radio conglomerate Clear Channel, which in 1995 operated 40 radio stations, today owns over 1,200 stations and controls 11 percent of the market.[17] Rupert Murdoch's News Corporation is the largest media conglomerate on the planet, one of seven media giants that own or control virtually all of the United States' 2,000 TV stations, 11,000 radio stations, and 11,000 newspapers and magazines.[18] And, predictably, these media corporations have the White House's support. Despite congressional mandates for diversity of ownership and local control, the number of corporations that control our media is shrinking dramatically.

This consolidation reduces diversity, gives consumers limited and homogenized choices, and erodes local control. Radio stations play the same music, giving little opportunity for new or alternative artists. North Dakota farmers can't get local emergency broadcasts or crop reports, and New York City residents no longer have a country radio station. Corporate consolidation has reduced news broadcast quality and has dramatically diminished the inquisitiveness of our national press.

To meet the challenges of the future, the United States needs an open marketplace of ideas. As fewer companies own more and more properties, that marketplace is withering. TV stations are no longer controlled by people primarily engaged in their communities, and news bureaus are no longer run by newspeople. Driven solely by the profit motive, many of these companies have liquidated their investigative journalism units, documentary teams, and foreign bureaus to shave expenses. Americans must now tune in to the BBC to get quality foreign news. Local news coverage is also shrinking, as owners cut corners by consolidating newsrooms. Coverage at the

Louisiana Statehouse in Baton Rouge is typical: In 1970 there were five investigative reporters assigned to the Capitol beat. Today there are none. Not a single reporter from a national news outlet is currently assigned to cover the U.S. Department of Interior.

I recently asked Fox News president Roger Ailes why the networks don't cover environmental stories. Roger is an old friend with whom I spent a summer camping in Africa almost 30 years ago. He is jovial, animated, and genuinely funny, and we loathe each other's politics. After considering the question for a moment, he said, "It's because environmental stories are not fast-breaking!" News, it seems, has to be entertaining because that's what sells.

The networks are contaminating the airwaves with high-profile murders and celebrity gossip, leaving ever-diminishing time for real news. They've dumbed down the news to its lowest common denominator. It's all Laci Peterson and Kobe Bryant all the time. Notorious crimes and sex scandals have little real relevance to our lives, our country, our democracy. At best, they are entertainment; at worst, pornography. The Monica Lewinsky story got such play in part because it was an excuse to deal pornography packaged as news. That stuff may sell papers, but it leaves little room for the asthma stories, for news that really affects our lives.

But Roger Ailes' response omitted another factor: Environmental stories often challenge a network's ideology or corporate self-interest. Many major media outlets are controlled by companies that have a vested interest in keeping environmental disasters under wraps: NBC is owned by General Electric, the world's biggest polluter, with a world record 86 Superfund sites.[19] Until three years ago, CBS was owned by Westinghouse, which has 39 Superfund sites. Westinghouse is also the world's largest owner of nuclear power plants and the third-largest manufacturer of nuclear weapons.[20]

In 2003, the North American winners of the prestigious Goldman Environmental Prize, known as the "Nobel Prize for grassroots work," were former Fox TV reporters Jane Akre and Steve Wilson. The two investigative reporters claim that they lost their jobs at Tampa's Fox-owned WTVT when they refused to doctor a news report that had displeased Monsanto.[21] The reporters had visited regional dairies and discovered that Monsanto's controversial bovine growth hormone (BGH) was being injected into cows by virtually every dairyman in the region.[22] The chemical was present in virtually all the state's milk supply, despite commitments by Florida's supermarkets not to sell milk tainted by the hormone.[23] In various studies BGH has been linked to cancer[24] and is banned by many countries, including Canada,[25] New Zealand,[26] and the entire European community.[27] Akre and Wilson's report said that Monsanto had been accused of fraud in connection with information it had provided to the EPA concerning dioxin, published deceitful statements about food safety, and funded favorable studies about the product from tame scientists.[28] The newscast also reported on allegations that Monsanto had attempted to bribe public officials in Canada.[29]

According to the reporters, WTVT carefully reviewed the team's four-part investigation for factual accuracy and heavily advertised the series on radio. It planned to release the story during television sweeps week beginning February 24, 1997. The day before the airing, however, the station yanked the shows after Monsanto hired a powerful law firm to complain to Roger Ailes.[30] Wilson and Akre testified that the local station manager again reviewed the reports, found no errors, and scheduled them to run the following week. The station also offered Monsanto an opportunity to appear on the show and respond. Monsanto declined the offer and fired off another threatening letter to Ailes. Wilson and Akre claim that the station manager, David Boylan, ordered the reporters to edit the show in a way that was deceptive but favorable to Monsanto.[31] "For every fact we intended to broadcast, we had documentation six weeks from Sunday," Wilson told me. "The station's lawyer told us time and again, 'You don't get it. It doesn't matter what the facts are, we don't want to be spending money to defend a lawsuit.'"[32] According to Wilson, the station was also worried about losing advertisers and had received calls from a grocery-chain and dairy-industry interests.

According to their subsequent lawsuit, Boylan threatened to fire Wilson and Akre "within 48 hours" if they declined to cooperate in the deception. He subsequently softened this position, they testified, offering to lay off both reporters with full salaries for their contract period, provided they agreed to sign a confidentiality agreement.[33] For nine months they worked on 83 different drafts of the story—none of which satisfied Fox or Monsanto. Akre testified that the station had tried to force her to say that the BGH milk was safe and no different from non-BGH milk, despite abundant studies that showed otherwise.[34] "We told them to go ahead and kill the story," Wilson says, "just don't make us lie."[35] Boylan eventually fired the reporters in December 1997, and they sued Fox. In August 2000, following a five-month trial, a Florida jury awarded Akre $425,000 under Florida's private-sector whistle-blower's statute, which prohibits retaliation against employees who threaten to disclose employer conduct that is "in violation of a law, rule or regulation."[36] The jury found that Akre had been fired "because she threatened to disclose to the Federal Communications Commission under oath in writing the broadcast of a false, distorted, or slanted news report that she reasonably believed would violate the prohibition against intentional fabrications or distortions of the news on television."[37]

But the story does not have an ending that is happy for Akre and Wilson, or for American democracy. On February 14, 2003, the Florida District Court of Appeals reversed the jury verdict. The bizarre decision adopted Fox's argument that the FCC's 50-year-old News Distortion Rule, which prohibits the broadcast of false reports, does not qualify as a "law, rule or regulation," as required by the whistle-blower's statute, since it had been created over the years in decisions by FCC judges and never promulgated in a rule-making process.[38]

Five major networks filed amicus curiae briefs supporting Fox's argument.[39] This decision effectively declared it legal for networks to lie in news reports to please their advertisers. Judge Patricia Kelly, the Jeb Bush–appointed district judge who wrote the opinion, next remanded the case to the trial court to determine whether Akre and Wilson should reimburse Fox for $1.7 million in legal fees.[40] The argument will take place in August 2004. "What reporter is going to challenge a network that orders him to cover up for polluters or companies that abuse workers or engage in health and safety violations if the station can retaliate by suing the reporter to oblivion the way the courts are letting them do to us?" asks Wilson.[41]

It should come as no surprise that a virtual media blackout greeted Akre and Wilson's reception of the Goldman Prize; their story has been largely ignored by the mainstream press. "The news today is far more about the business of journalism than the journalism business," Akre complained to me.[42] Wilson observed that "if you own a newspaper or a printing press, you can lie to your heart's content. But if you are using the public airwaves, you have an obligation to be fair, accurate, and truthful, even in circumstances where it's going to piss off your advertisers, embarrass your friends, or hurt your bottom line—otherwise you're violating the public trust and stealing something vital from the public."[43]

Not long ago, people scoffed at the suggestion that a network's corporate owner would censor news out of self-interest. That can't happen in America, right? But times have changed. Everybody saw how CBS genuflected to the right wing and the Republican National Committee to pull a docudrama that was critical of Ronald Reagan. (CBS's hypervigilance, of course, did not apply to Janet Jackson's naked breast.) The Reagan show was tasteless and historically inaccurate, but that's never stopped CBS from airing similar shows about other prominent political figures. Despite having the highest-rated show on his network, Phil Donahue got sacked by MSNBC because of his liberal philosophy. MSNBC replaced him with a right-wing bigot, Michael Savage.

The corporate bias infects nearly every major news outlet. Michael Eisner has said that he doesn't want ABC News to report critically on Disney, its parent company. In May 2004, Eisner canceled distribution of Michael Moore's *Fahrenheit 9/11*—a screed against George W. Bush. According to Moore's agent, Ari Emmanuel, Eisner feared that Governor Jeb Bush would rescind tax breaks now granted to the company's Florida theme parks. What about behemoths like GE, which has subsidiaries with financial stakes in myriad public policy debates from war to pollution?

I have considerable personal experience with corporate censorship. Charles Grodin often reminds me that I got him fired from the best job he ever had—as a nightly talk-show host on MSNBC.

On November 11, 1996, Grodin had me on his show to plug my book *Riverkeepers.* Unlike the more seasoned MSNBC and NBC hosts, he allowed me to talk at length about the record of the network's parent company, GE. I talked about GE's massive pollution of the Hudson River, about the fact that GE owns more Superfund sites than any other company, and that, thanks to GE pollution, hundreds of fishermen were now jobless, while then-CEO Jack Welch took home an $85 million salary plus bonuses.

A few months later his bosses canceled the show so suddenly that Grodin didn't even get to say good-bye. In a postmortem column, New York *Newsday* journalist Marvin Kitman mourned the surprise sacking of Grodin, which he attributed to my interview. Kitman commented that my appearance "was the longest attack on a General Electric–owned network on GE for polluting the Hudson" and lamented that Grodin "was one of the things that was good about TV, a genuine original, the closest thing we had to an Oscar Levant in this age of mellow-mouth talk-show hosts."[44] According to Grodin, Ralph Nader called Jack Welch to protest the sacking, but Welch never returned the call.

I regularly run afoul of corporate censors and bean counters who decide television content. In November 2003, when environmentalists around the country were engaged in fighting the Cheney energy bill, the NRDC was anxiously trying to get me airtime because no one was talking about the bill on TV. Fox TV host Bill O'Reilly agreed to schedule me, but only with the explicit proviso that I wouldn't say critical things about George W. Bush. I would first have to do a pre-interview to make sure I was capable of talking about the environment without badmouthing the president. Later, Fox decided that even this was too chancy; they would just tape the show, rather than risk me going off the reservation on live TV. The same week, Tom

Brokaw, a committed environmentalist and fly fisherman, scheduled me for a segment on *NBC Nightly News*—but the producers bumped me for yet another Michael Jackson story.

I was most disappointed by Aaron Brown of CNN. When Ted Turner owned the network, CNN was a bastion of environmental reporting in the wasteland of network news shows. Turner employed an environmental specialist, Barbara Pyle, as a full-time advocate for environmental programming. But CNN then became an AOL Time Warner property, and on the day I was scheduled to appear, one of Brown's producers called to cancel the interview. Brown, she said, was aware of my criticism of the president's environmental record and was canceling my appearance because he didn't want any "Bush bashing" on his show. Brown, too, substituted the interview with me for a segment on Michael Jackson's sex scandal.

I was at the National Press Club in Washington, D.C., the next morning to give a speech. As I waited for the elevator, I read the Journalist's Creed from the plaque in the foyer:

> I believe in the profession of journalism. I believe that the public journal is a public trust; that all connected with it are, to the full measure of responsibility, trustees for the public; that acceptance of lesser service than the public service is a betrayal of this trust; that individual responsibility may not be escaped by pleading another's instructions or another's dividends; that advertising, news and editorial columns should alike serve the best interests of readers; that supreme test of good journalism is the measure of its public service.

Sleazy scoundrels like Steven Griles and Jeffrey Holmstead or medicine-show fakirs like John Graham make the endlessly broadcast Clinton-Whitewater scandal look like a Sunday-school romp, yet they are invisible in the press. "The networks are owned by big corporations and they're mainly Republican," DNC chairman Terry McCauliffe recently complained to me. "It's a heavy lift getting them to cover corporate control issues or to criticize a Republican president."

Public interest advocates can't criticize corporations on the airwaves even when they have the money. Moveon.org learned this lesson when they tried unsuccessfully to air an ad criticizing President Bush's corporate coddling during the Super Bowl. In 2003, when Laurie David and Arianna Huffington's "Detroit Project" attempted to air paid advertisements touting automobile fuel efficiency, the networks, which make $15 billion annually from the auto industry, refused to carry the ads. "They wouldn't run them," Huffington told me. "And we ended basically not being able to use the money that was budgeted to buy airtime."[45] Huffington turned to Laurie David, a former David Letterman producer whose husband, Larry David, created *Seinfeld* and the popular series *Curb Your Enthusiasm.* "I met with Lloyd Braun, the president of ABC," David told me, "and brought the commercial up there to see if they could run the ads. He pretty much laughed me out of the office. He said, 'We have three offices. We have an office in Los Angeles, we have an office in New York City, and our third office is in Detroit.' There was no way he was going to put something on his network that might piss off the auto industry."[46]

When George W. Bush arrived at the White House, there was still one significant media law in place: No media company was allowed to dominate any one particular market. But Bush's FCC is looking sideways while the media giants violate this restriction. FCC regulations prohibit ownership of more than 8 radio stations in a single market. A recent study of 337 cities by the Center for Public Integrity found giant corporations owning more than 8 stations in 34 of them.[47] Clear Channel is the big kahuna, with 11 of the 17 radio stations in Mansfield, Ohio.[48] Second in size after Clear Channel is right-wing Cumulus Media, which enforced skinhead-style censorship when it blackballed the Dixie Chicks for criticizing President Bush. Cumulus owns 8 of the 15 stations in Albany, Georgia.[49] In every city surveyed, a single company owns at least one-third of the radio outlets.[50]

The TV companies are engaged in the same shenanigans. The FCC rule that forbids ownership of more than one TV station in any market has been broken in 43 cities surveyed by the Center for Public Integrity. Recently, for example, Fox's affiliate in Wilmington, North Carolina, was purchased by a company that turns out to be a sister subsidiary of the company that already owns the NBC affiliate. They fired staff and combined newsrooms, so now one media company controls two of Wilmington's three stations.[51] When Rupert Murdoch's News Corporation bought Chris Craft's TV stations and Viacom merged with CBS in 2000, both companies were suddenly violating FCC rules prohibiting a single entity from owning stations reaching over 35 percent of the national audience. The FCC, now chaired by merger-maniac Michael Powell, solved the problem by handing both companies temporary waivers.[52] Then the FCC tried to make the waivers permanent by raising the limit on market share.[53] This new FCC rollback will unleash the largest wave of media consolidation in U.S. history. The new rules allow gigantic media conglomerates to buy television stations reaching 45 percent of the nation's viewers and to own newspaper, radio, and television stations in the same city.

Chairman Powell, Secretary of State Colin Powell's son, conducted his rule-making proceedings in virtual secrecy, confining debate to a single public hearing in Richmond, Virginia, on February 27, 2003. Not surprisingly, it received very little attention from the TV networks. The big newspaper chains—the *New York Times,* Knight Ridder, and Gannett—enjoying their own unprecedented consolidations and creating their own plans to enter the television market—all but blacked out coverage as well. In June 2003, Powell and his two Republican commissioners announced the deal as a fait accompli.

But Powell's corporate sop ignited a firestorm as conservatives, frightened by the prospect of monolithic corporate control of the nation's fundamental freedom, joined liberals in protest. Senator John McCain pointed out that a similar media consolidation had subverted Russia's new democracy. Conservative columnist William Safire campaigned in favor of bipartisan legislation in the Senate to kill the deal. Public pressure forced Powell to reopen the process and hold open meetings in cities across the United States. A record 2.4 million people wrote letters opposing the rollbacks,[54] recognizing what George W. Bush and Michael Powell apparently do not—that the control of our media by a half-dozen powerful multinationals who can dictate what we hear, see, and read is dangerous for our communities, our families, and our democracy.

The Senate voted to stop the deal, and the House had sufficient votes to do the same. But the White House, working with Tom DeLay, the media moguls, and their lobbyists, blocked the vote. Fortunately, in June the federal court of appeals in Philadelphia rejected the FCC's rollbacks, citing a lack of "reasoned analysis," and directed the agency to start over.[55]

Nevertheless, absent a resurrection of the Fairness Doctrine, our nation's broadcast media, which should be an open forum for our democracy, will continue to devolve into a marketplace exclusively for commerce. It allows these corporations to extend the reach of their empires into American homes with customized, interactive multimedia content hell-bent on transforming us into 24-hour-a-day consumers. The so-called news and entertainment content will be dictated by advertisers with personalized appeals calculated to program us to buy, buy, buy. Meanwhile, our civic life, already invisible on TV, will become an irrelevant relic to the next generation, which will know little about the issues or why they should participate in democracy.

11

Reclaiming America

You show me a polluter and I'll show you a subsidy. I'll show you a fat cat using political clout to escape the discipline of the free market and load his production costs onto the backs of the public.

The fact is, free-market capitalism is the best thing that could happen to our environment, our economy, our country. Simply put, true free-market capitalism, in which businesses pay all the costs of bringing their products to market, is the most efficient and democratic way of distributing the goods of the land—and the surest way to eliminate pollution. Free markets, when allowed to function, properly value raw materials and encourage producers to eliminate waste—pollution—by reducing, reusing, and recycling.

As Jim Hightower likes to say, "The free market is a great thing—we should try it some time."

In a real free-market economy, when you make yourself rich, you enrich your community. But polluters make themselves rich by making everybody else poor. They raise the standard of living for themselves by lowering the quality of life for everyone else. And they do that by escaping the discipline of the free market.

The coal-burning utilities that acidify the Adirondack lakes, poison our waterways with mercury, provoke 120,000 asthma attacks, and kill 30,000 of our neighbors every year are imposing costs on the rest of us that should, in a free-market economy, be reflected in the price of the energy when they bring it to the marketplace. By avoiding these costs, the utilities are able to enrich their shareholders and put their more conscientious and efficient competitors out of business. But these costs don't disappear. The American people pay for them downstream—with poisoned fish, sickened children, and a diminished quality of life. Every one of our federal environmental laws is intended to restore true free-market capitalism so that the price of bringing a product to market reflects the costs that it imposes on the public.

The truth is, I don't even think of myself as an environmentalist anymore. I consider myself a free-marketeer. Along with my colleagues at the NRDC and Waterkeeper, I go out into the marketplace and catch the cheaters. We tell them, "We're going to force you to internalize your costs the same as you internalize your profits." Because when polluters cheat, it distorts the entire marketplace, and none of us benefits from the efficiencies and democracy that the free market promises.

Corporate capitalists don't want free markets, they want dependable profits, and their surest route is to crush the competition by controlling the government. The domination of our government by large corporations leads to the elimination of markets and, ultimately, to the loss of democracy.

Some of the largest federal subsidies are going to western

resource industries—grazing, lumber, mining, and agribusiness—that have spawned the most vocal attacks against federal environmental laws. These industries are run by some of the richest and most radically conservative people in the country, men like Richard Mellon Scaife, Charles Koch, and Joseph Coors. Their intense hatred for federal government is, in a supreme irony, combined with an intense reliance on federal subsidies. Let's not forget that we taxpayers give away $65 billion every year in subsidies to big oil, and more than $35 billion a year in subsidies to western welfare cowboys, many of whom are destroying our public lands and waterways. Those subsidies helped create the billionaires who financed the right-wing revolution on Capitol Hill and put George W. Bush in the White House. And now they have indentured servants in Washington demanding that we have capitalism for the poor and socialism for the rich.

The free market has been all but eliminated in an energy sector dominated by cartels and monopolies and distorted by obscene subsidies to the filthiest polluters. Our once vibrant agricultural markets are now controlled by multinational monopolies with no demonstrated loyalty to our country or its laws. Media consolidation is transforming journalism from a forum of ideas into a marketplace exclusively for commerce.

If you haven't already done so, say good-bye to the merchants who anchor our local economies and communities. While profits from the big-box stores flow to distant corporate headquarters, struggling small businesses and farmers recycle their profits back into their communities through their support of Boy Scouts, Little Leagues, and Rotary Clubs, through local commerce, and by paying local employees a living wage and benefits. They pay taxes (a duty shirked by 61 percent of large corporations),[1] and they don't move their corporate headquarters to Bermuda and their operations to Taiwan. These local entrepreneurs are the training schools for civic leadership, and the loss of them sounds the death knell for consumer choice, civic life, and community investment.

Teddy Roosevelt often observed that American democracy is too sturdy to be destroyed by a foreign enemy. But, he warned, it could easily be destroyed by "malefactors of great wealth" who would subvert our political institutions from within.[2]

Roosevelt was no isolated Cassandra. Our greatest political icons from Thomas Jefferson onward have warned Americans against allowing corporate power to dominate our political landscape. In his most famous speech, President Dwight Eisenhower cautioned Americans about the grave danger of falling under control of "the military-industrial complex."[3] In 1863, in the depths of the Civil War, Abraham Lincoln is said to have lamented, "I have the Confederacy before me and the bankers behind me, and for my country I fear the bankers most." Franklin Roosevelt echoed that sentiment when he warned that "the liberty of a democracy is not safe if the people tolerate the growth of private power to a point where it becomes stronger than their democratic state itself. That, in its essence, is fascism."[4]

While communism is the control of business by government, fascism is the control of government by business. My *American Heritage Dictionary* defines fascism as "a system of government that exercises a dictatorship of the extreme right, typically through the merging of state and business leadership together with belligerent nationalism." Sound familiar?

The rise of fascism across Europe in the 1930s offers plenty of lessons on how corporate power can undermine democracy. While the United States confronted its devastating depression by reaffirming its democracy—enacting mininum wage and Social Security laws to foster a middle class, passing income taxes and antitrust legislation to limit the power of corporations and the wealthy, and commissioning parks and public lands and museums to create employment and safeguard the commons—Spain, Germany, and Italy reacted to their economic crises in a very different manner. Industrialists forged unholy alliances with right-wing radicals and their charismatic leaders to win elections in Italy and Germany, and then flooded the ministries, running them for their own profit, pouring government money into corporate coffers, and awarding lucrative contracts to prosecute wars and build infrastructure. Benito Mussolini's inside view of the process led him to complain that "fascism should more appropriately be called 'corporatism' because it is the merger of state and corporate power."

These elected governments used the provocation of terrorist attacks, continual wars, and invocations of patriotism and homeland security to privatize the commons, tame the press, muzzle criticism by opponents, and turn government over to corporate control. "It is always a simple matter to drag the people along," noted Hitler's sidekick, Hermann Goering, "whether it is a democracy, or a fascist dictatorship, or a parliament, or a communist dictatorship. Voice or no voice, the people can always be brought to the bidding of the leaders. That is easy. All you have to do is tell them they are being attacked, and denounce the peacemakers for lack of patriotism and exposing the country to danger. It works the same in any country."

The White House has clearly grasped the lesson. The Bush

administration won't ask its industry paymasters to protect their chemical and nuclear plants, but instead has devised an alert system seemingly designed to keep Americans in a constant state of apprehension. As to the war on terror, "It may never end," warned Vice President Cheney in October 2001. "At least, not in our lifetime."

The historian Alex Carey observed that the twentieth century has been largely shaped by three trends: "The growth of democracy, the growth of corporate power, and the growth of corporate propaganda as a means of protecting corporate power against democracy." The Bush administration marks the triumph of this last trend. Under George W. Bush, American government is dominated by corporate power to an extent unprecedented since the Gilded Age, when the sugar, oil, steel, and railroad trusts owned government officials and traded them like commodities.

While claiming to embrace its values, the Bush administration has stolen the soul of the Republican Party. The president and his cronies have taken the conserve out of conservative. Instead of rugged individualism, they've created a clubhouse that dispenses no-bid contracts to Halliburton. They talk about law and order while encouraging corporate polluters to violate the law. They proclaim free markets while advocating corporate welfare. They claim to love democracy while undermining open government. They applaud state rights and local control, but they are the first to tear up local zoning laws and bully states into lowering environmental standards to make way for corporate profit taking. They exalt property rights, but only when it's the right of a property owner to use his property to pollute or destroy someone else's. Where are these property rights advocates when big coal is demolishing homes in Appalachia, when coal-bed methane barons are destroying Wyoming ranches, when the hog barons are defiling property in North Carolina?

While condemning environmentalists as "radicals," they promote the radical notion that clean water, clean air, and healthy loved ones are luxuries we can't afford.

They invoke Christianity to justify the rape of the land, violating manifold Christian precepts that require us to be careful stewards. Rather than elevating the human spirit, their interpretation of Scripture emphasizes the grimmest vision of the human condition. They embrace intolerance, selfishness, pride, arrogance toward creation, and irresponsibility to the community and future generations.

The easiest thing for a political leader to do is appeal to our fear, our hatred, our greed, our prejudices. My most poignant memory of my father came in the days after he died. I was 14 at the time. After his wake at St. Patrick's Cathedral in New York City, we took him on the train to Washington, D.C. I will never forget the hundreds of thousands of people who lined the tracks—blacks, whites, priests, nuns, rabbis, hippies, men in uniform—many with tears running down their faces, many waving American flags or carrying signs bidding "Good-bye Bobby." From Union Station in Washington we rode in a convoy past the Mall, where thousands of homeless men were encamped in shanties left from Martin Luther King Jr.'s last campaign, and they came to the edge of street and stood with their heads bowed as we passed, crossing the Potomac and heading up the hill at Arlington to bury my father under a simple stone next to his brother.

The faces I saw that day were a cross section of America, the faces of the American community. And yet four years later, I learned from polling data that many of the white people who had supported my father in the 1968 Maryland primary, and had then waved to us from beside that railway track, had voted not for George McGovern in the 1972 primary, but for George Wallace, a man whose philosophies were diametrically opposed to everything my father stood for. It struck me then—and my observations have confirmed this many times since—that every nation, and every individual, has a dark side and a light side. And the simplest strategy for a politician is to exploit our baser instincts.

Our greatest politicians have accepted the tougher task of appealing to our sense of community, asking Americans to transcend their own self-interest. Throughout our history they have persuaded us to find the hero in ourselves, and to make sacrifices on behalf of future generations—and for the principles that underpin America's unique mission. John Winthrop, the Moses of the Puritan migration, said that mission was to build a "city on a hill"—an example to the world of what nations can accomplish if we work together in community. Winthrop's 1630 sermon—arguably the most important speech in American history—called for his fellow citizens to steer away from the greed and power politics that had corrupted the old-world culture. He urged people to build a land that would be "a model for Christian charity." Winthrop's words are often quoted by neoconservatives who invariably omit his warning against the temptation to elevate commercial values lest we "disappear into the lure of real estate."

But instead of inspiring us with invocations for courage, community, and sacrifice, President Bush's campaign strategy revolves around fear-mongering and appeals to selfishness—Karl Rove's two *t*'s: taxes and terrorists. Instead of can-do American ingenuity, this is the administration of "can't do." It has constructed a philosophy of government based on self-interest run riot: It has borrowed $9 trillion from our children and looted our Treasury, poisoned our water and air, destroyed our public lands, and sacrificed our health—all to enrich the wealthy few. It has reduced the honorable profession of public service to an opportunity for plunder and self-enrichment.

In *The Shame of the Cities,* his watershed 1904 study of the American political system, Lincoln Steffens concluded that the corruption and failures of American democracy stemmed largely from a single source—the control of government by businesspeople acting in their own self-interest. Steffens characterized that formula as a kind of treason because "the effect of it is literally to change the form of government from one that is representative of the people to an oligarchy representative of the special interests."

Generations of Americans will pay for the Republican campaign debt to the energy industry and other big polluters with global instability, depleted national coffers, and increased vulnerability to oil-market price shocks. They will also pay with reduced prosperity and quality of life at home. Pollution from power plants and traffic smog will continue to skyrocket. Carbon dioxide emissions will aggravate global warming. Acid rain and mercury will continue to sterilize our lakes, poison our fish, and sicken our people. The administration's attacks on science and the law have put something perhaps even greater at risk—our values and our democracy.

George W. Bush and his court are treating our country as a grab bag for the robber barons, doling out the commons to giant polluters. Together they are cashing in our air, water, aquifers, wildlife, and public lands and divvying up the loot. They are turning our politicians into indentured servants who repay campaign contributions with taxpayer-funded subsidies and lucrative contracts and reign in law enforcement against a booming corporate crime wave.

If they knew the truth, most Americans would share my fury that this president is allowing his corporate cronies to steal America from our children.

Acknowledgments

I want to thank Jann Wenner, who convinced me to summarize George W. Bush's environmental offenses for a *Rolling Stone* article published in December 2003 and who conceived the title of this book. Thanks also to my lifelong friend Peter Kaplan, who steered me to my editor, Mark Bryant, at HarperCollins, persuaded me to write this book, and then read and critiqued the manuscript.

Many people worked double-time to help me get this book out quickly. In particular, I owe thanks to:

Mary Beth Postman, who worked around the clock to carefully coordinate a baffling array of drafts and footnotes. I am grateful for her motivational skills and good judgment.

Lori Morash transferred my long-hand chicken scratch into print at lightning speed, often laboring on nights, weekends, and holidays.

David Ludlum coordinated fact-checking for not only this book but also an article in *Rolling Stone* and another in *The Nation,* both of which inspired this book. I am grateful for his hard work, care, thoroughness, and commitment, not to mention his willingness to relinquish a long-planned Texas vacation to work on the endnotes.

I'm indebted to my superb editors: Delia Marshall, who, with unerring judgment, carved my 100,000-word manuscript in half and reorganized the chapters into a sensible configuration; and Mark Bryant, who helped me polish the book with brilliant editorial recommendations. Mark's confidence in the book, and our long conceptual conversations about it, made a difficult task fun.

My agent, Kris Dahl of ICM, proved herself, as always, a wise and extraordinary resource, and a wonderful friend.

My thanks to Pace Law students Stephanie Haggerty, Audrey Friedrichsen, Elana Roffman, Erin Flanagan, Daniel Yohannes, Jennifer Nelson, Robert Manfredo; to Riverkeeper interns Daniel Jacobson, Garen McClure; to EPA attorney Marla Wieder; and to former EPA attorney Janet MacGillivray, who helped me research and then thoroughly fact-checked the book.

Waterkeeper Alliance's attorneys Kevin Madonna and Daniel Estrin read and fact-checked my sections on industrial hog production and the Wise Use Movement.

Riverkeeper's Reed Super advised me on power plant issues, and Kyle Rabin, also at Riverkeeper and a human encyclopedia on nuclear power plants, helped me understand the vulnerabilities of those plants as a national security issue.

To all my colleagues at the Natural Resources Defense Council who provided research, insight, and support on different parts of the book, especially Greg Wetstone, Wesley Warren, Dave Hawkins, Erik Olson, Rob Perks, Jon Coifman, Jon Devine, Daniel Rosenberg, Karen Wayland, Sharon Buccino, John Walke, Aaron Colangelo, Linda Greer, Jennifer Sass, Dan Lashof, Johanna Wald, Chuck Clusen, Alyssondra Campaigne, Alan Metrick, Melanie Shepherdson, Tom Cochran, Andrew Wetzler, and Kidd Dorn.

Kristin Sykes; Joan Mulhern of Earthjustice; and Louise Dunlap, formerly of the Environmental Policy Institute, contributed research and advice about Steven Griles and mountaintop mining. Attorney Ed Gramdis and Joe Lovett of the Appalachian Center for the Environment and the Economy were also a helpful resources on mountaintop mining. Charlene Epperson helped me locate and interview Interior Department insiders on the subject.

My friend and fishing buddy Dick Russell came to my aid on the Texas chapter at the moment I most needed help. I am also indebted to Molly Ivins and Lou Dubose for their research on George W. Bush's environmental record as governor.

I also want to thank OMB Watch and its former policy analyst Reese Rushing for research on John Graham's record, and Public Citizen for the research that it maintains in its online "Graham Watch." I am grateful to the Center for Media & Democracy and its "PR Watch," and to Bill Moyers, whose thoughtful work has helped alert America to the threat that current trends in media consolidation pose to our democracy. The Center for Responsible Politics provided invaluable help with its online database of campaign contributors, as did Common Cause with its report "Chemical Reaction."

I am most indebted to my children—Bobby, Kick, Conor, Kyra, Finbar, and Aidan, who uncomplainingly put up with weekends without their dad and who gave up our regular ski trips to the Catskills so that I could grind away at the book; to my brilliant partner Karl Coplan, who covered my cases and classes so that I could focus on the writing; and to my friends Laurie David and Mike Papantonio, without whose support and encouragement I could not have undertaken this project.

Notes

NOTE: Internet sources are supplied case-sensitive.

Introduction

1. The Luntz Companies, "The Environment: A Cleaner, Healthier, Safer America," viewed at http://www.ewg.org/briefings/luntzmemo/pdf/LuntzResearch_environment.pdf.
2. "Key Facts: Race, Ethnicity and Medical Care," The Henry J. Kaiser Family Foundation, June 2003, viewed at http://www.kff.org/minorityhealth/6069-index.cfm.
3. Janet MacGillivray, interview by Robert F. Kennedy, Jr., June 2004.
4. "The New Nationalism," *National Edition of the Works of Theodore Roosevelt,* vol. XVII (New York, 1910), p. 52.

1| The Mess in Texas

1. Tom Horton, "Texas Economy: George W. Bush and the Environment," *Rolling Stone,* December 11, 2003.
2. *Water Quality Program and Assessment Summary,* Texas Natural Resources Conservation Commission, September 2003, pp. 8-16, available at http://www.tnrcc.state.tx.us/water/quality/02_twqmar/02_305b/02_program_summary/13-str&rivass.pdf

3. *Texas 2000 Air Quality Study,* Texas Natural Resources Conservation Commission, August-September 2000, p. 1, available at http://www.utexas.edu/research/ceer/texaqs/visitors/news1.pdf.
4. "George W. Bush: The Polluters' Governor," Sierra Club, February 16, 2000, available at http://www.commondreams.org/news2000/0216-01.htm. The Sierra Club based its findings on its study of the EPA's Toxics Release Inventory Database, 1990-1997, and Emergency Response Notification System Database.
5. *Remember the Past, Protect the Future,* U.S. EPA, Region 6, April 2000, p. 30.
6. "Polluters Bet Big on Bush on the Campaign Money Trail," report by Texas PEER, November 1995, viewed at http://www.txpeer.org/Bush/Polluters_Bet_On_Bush.html.
7. Frederick Lurman et al., "Assessment of the Health Benefits of Improving Air Quality in Houston, Texas," Sonoma Technology, Inc., November 1999, p. ES-4.
8. Tom "Smitty" Smith, interview by Dick Russell, February 20, 2004; Texas PEER, "Polluters Bet Big."
9. "Bush's Environmental Record," *NewsHour with Jim Lehrer* transcript, August 22, 2000, viewed at http://www.pbs.org/newshour/bb/election/july-dec00/bush_environment_8-22.html.
10. Louis Dubose, "Running on Empty," *The Nation,* April 26, 1999.
11. Vulliamy, "Dark Heart of the American Dream," available at http://observer.guardian.co.uk/magazine/story/0,11913,738196,00.html; Bob Herbert, "Bush Goes Green," *New York Times,* April 6, 2000; Jay Root, "Bush Record on Pollution in Texas Draws Fire," *Chicago Tribune,* November 26, 1999; Molly Ivins and Louis Dubose, "Bush and the Texas Environment," *Texas Observer,* April 14, 2000, available at http://www.bushfiles.com/00_04_14/000414_bush_and_environment.htm.
12. Herbert, "Bush Goes Green."
13. Ralph K. M. Haurwitz and Stuart Eskenazi, "Statement Equating Slaves with Property Denounced," *Austin-American Statesman,* October 14, 1994. Kuykendall made the reference twice in September 1994: first at a forum in Kerrville sponsored by the Kerr County Republicans Club and subsequently during a

call-in session on KUT-FM, the public radio station of the University of Texas.

14. Texas PEER, "Polluters Bet Big."
15. Molly Ivins and Louis Dubose, "Bush and the Texas Environment."
16. Neil Carman, interview by Dick Russell, February 20, 2004.
17. Molly Ivins and Louis Dubose, *Shrub: The Short but Happy Political Life of George W. Bush* (New York: Random House, 2000), p. 119.
18. John Coequyt, "Fuzzy Air: Why Texas is the Smoggiest State," Environmental Working Group Policy Analysis, October 17, 2000, available at http://www.ewg.org/reports_content/fuzzy air/fuzzyair.pdf.
19. Horton, "Texas Economy."
20. Natural Resources Defense Council Press Release, "U.S. Beach Closings at Record High for Second Year in a Row Despite Drought, According to Annual NRDC Beach Quality Report," August 3, 2000, available at http://www.nrdc.org/media/press Releases/000803b.asp.
21. Michael King, "A Thousand Points of Darkness," *Austin Chronicle,* January 12, 2001; Vulliamy, "Dark Heart of the American Dream."
22. Rose Farley, "Bottom of the Ninth," *Dallas Observer,* February 12, 1998, viewed at http://www.dallasobserver.com/issues/1998-02-12/news2.html.
23. Rose Farley, "Something In The Air," *Dallas Observer,* June 19, 1997, viewed at http://www.dallasobserver.com/issues/1997 06-19/feature.html.
24. Ibid.
25. Jim Yardley, "Bush Approach to Pollution: Preference for Self-Policing," *New York Times*, November 9, 1999.
26. Texas PEER, "Polluters Bet Big."
27. Neil Carman, interview by Dick Russell, February 20, 2004.
28. Vulliamy, "Dark Heart of the American Dream.
29. "Texas Air Quality Study 2000," viewed at http://www.utexas.edu/research/ceer/texaqs.
30. Herbert, "Bush Goes Green."

31. Texas PEER, "Polluters Bet Big."
32. Yardley, "Bush Approach to Pollution."
33. Jeffrey St. Clair, "Cash and Carry," *In These Times,* March 6, 2000.
34. Janet Hook, "GOP's Go-To Guy Could Pose Risks for President," *Los Angeles Times,* June 16, 2003, p. A1.
35. George W. Bush, *A Charge to Keep* (New York: William Morrow, 1999).

2| Back to the Dark Ages

1. J. I. Bregman, "How to Clean Up the Mess," *World and I,* June 1989, available at http://www.worldandi.com/specialreport/1989/june/Sa16169.htm. No known attempt to prevent air pollution was made until the opening of the fourteenth century, when an antismoke ordinance forbidding the use of "sea coal" in London was established by royal proclamation. It is believed that at least one violator of this law was put to death by order of Edward I.
2. Gwen Florio, "Coors Has a Reputation, but No Political Record to Run On," *Rocky Mountain News,* April 10, 2004.
3. Russ Bellant, *The Coors Connection* (Boston: South End Press, 1991), pp. 85-87; Carl Deal, *The Greenpeace Guide to Anti-Environmental Organizations* (Berkeley, California: Odonian Press, 1993), p. 64; John Stauber and Sheldon Rampton, *Toxic Sludge Is Good for You!: Lies, Damn Lies and the Public Relations Industry* (Monroe, Maine: Common Courage Press, 1995), p. 142.
4. "Heritage Foundation President Mourns Death of Joseph Coors," Heritage Foundation press release, March 16, 2003.
5. David Helvarg, *The War Against the Greens* (San Francisco: Sierra Club Books, 1994), p. 20; Deal, *The Greenpeace Guide,* p. 58.
6. See "The Heritage Foundation," Anarchy for Anybody, available at http://www.cat.org.au/a4a/fake25.html.
7. People for the American Way, "Follow the Money: Funding and Support for Voucher Programs," viewed at http://www.pfaw.org/pfaw/general/default.aspx?oid=10723; Capital Re-

search Center database, viewed at http://www.capitalresearch.org/search/orgdisplay.asp?Org=HER100#fin.

8. John Saloma III, *Ominous Politics: The New Conservative Labyrinth* (New York: Hill and Wang, 1984), p. 14.
9. Bellant, *The Coors Connection,* pp. 9-10.
10. Ibid., p. 87.
11. Helvarg, *The War Against the Greens,* p. 69.
12. Ibid.
13. Dale Russakoff, "James Watt and the Wages of Influence," *Washington Post,* May 4, 1989, p. A21.
14. Lawrence Mosher, "Move Over, Jim Watt, Anne Gorsuch Is the Latest Target of Environmentalists," *National Journal,* October 24, 1981, p. 1899.
15. Ibid.
16. Tom Turner, Friends of the Earth, interview by David Ludlum, May 6, 2004.
17. Michael J. Sniffen, "Watt Charged with Coverup in HUD Scandal," Associated Press, February 22, 1995.
18. Helvarg, *The War Against the Greens,* p. 72.
19. Ibid.
20. Ibid., pp. 76-77.
21. Katherine Long, "His Goal: Destroy Environmentalism—Man and Group Prefer That People Exploit the Earth," *Seattle Times,* December 2, 1991.
22. Helvarg, *The War Against the Greens,* pp. 260-61.
23. Michael Lind, "Rev. Robertson's Grand International Conspiracy Theory," *New York Review of Books,* February 2, 1995.
24. Esther Diskin, "Robertson's Book Parallels Militia's Ideas," *The Virginian-Pilot* [Norfolk], April 30, 1995, p. A1.
25. Stauber and Rampton, *Toxic Sludge Is Good for You!,* pp. 85-87.
26. Alan Cooperman, "DeLay Criticized for 'Only Christianity' Remarks," *Washington Post,* April 20, 2002, p. A05.
27. Robert F. Kennedy, Jr., *The Riverkeepers* (New York: Scribner, 1997), p. 230.
28. David Greenberg, "The Clinton Warrior," *Washington Monthly,* June 1, 2003.
29. "Evolution Revolution," PBS.org., viewed at http://www.pbs.

org/wgbh/evolution/religion/revolution/1990.html.

30. Michael Weisskopf and David Maraniss, "Forging an Alliance for Deregulation," *Washington Post,* March 12, 1995, p. A1.
31. David Rogers, "General Newt: GOP's Rare Year Owes Much to How Gingrich Disciplined the House," *Wall Street Journal,* December 18, 1995.
32. Pat Williams, University of Montana Center for the Rocky Mountain West, interview by David Ludlum, February 26, 2004.
33. Ron Arnold, Center for the Defense of Free Enterprise, interview by David Ludlum, February 24, 2004.
34. Sheila Callahan, National Legal Center for the Public Interest, interview by David Ludlum, February 24, 2004.
35. Maria Weidner and Nancy Watzman, *Paybacks, Policy, Patrons, and Personnel: How the Bush Adminsitration Is Giving Away Our Environment to Corporate Contributors,* Earthjustice and Public Campaign, 2002, viewed at http://www.publicampaign.org/publications/studies/paybacks/Paybacks.pdf.
36. The Center for American Progress & OMB Watch for the Citizens for Sensible Safeguards Coalition, "Special Interest Takeover: The Bush Administration and the Dismantling of Public Safeguards," May 25, 2004, available at http://www.sensible safeguards.org/sit.phtml.
37. "Rewriting the Rules: The Bush Administration's Assault on the Environment," NRDC, April 2004, p. 3, viewed at http://www.nrdc.org/legislation/rollbacks/rr2004.pdf.
38. Ibid, p. 1.
39. Andrew Martin, "Documents Detail Lobbyists' Impact on Air-Quality Plan," *Chicago Tribune,* May 16, 2004.
40. See Mackinac Sierra Club website, http://michigan.sierraclub.org/issues/cafos/hazards.html.
41. Eric Schaeffer, former director of the EPA's Office of Regulatory Enforcement, interview by Robert F. Kennedy, Jr., September 2004.
42. "Koch Industries Indicted for Environmental Crimes at Refinery," U.S. Department of Justice press release, September 28, 2000, viewed at http://www.usdoj.gov/opa/pr/2000/September/

573enrd.htm; James Pinkerton, "Koch Slapped with Big Penalty/Guilty of Hiding Pollution Violation," *Houston Chronicle,* April 10, 2001.

43. NRDC, "The Bush-Cheney Energy Plan," viewed at http://www.nrdc.org/air/energy/aplayers.asp.
44. "Ron Arnold named as ecoterrorism expert in major study," Center for the Defense of Free Enterprise press release, viewed at http://www.cdfe.org/arnold_named_as_consultant.htm.
45. Paul Krugman, "Noonday in the Shade," *New York Times,* June 22, 2004.
46. In contrast, the environmental "monkey-wrenchers" Earth First and Earth Liberation Front (ELF) ascribe to a strict coda to protect all animal and human life. For example, when ELF activists burned Hummers at a California car dealership, they first entered the shop to move aquariums of live fish to safety. In contrast, anti-abortion terrorists have murdered 11 Americans and caused millions of dollars of property damage. Right-wing terrorists are responsible for the Oklahoma City bombing, which killed 168 people, and countless other attacks and acts of terrorist violence against American citizens (see David Helvarg, *The War Against the Greens;* Robert L. Snow, *Terrorists Among Us: The Militia Threat* [New York: Perseus Books Group, 2002]). The Bush administration has taken pains to distract the public from the terrorist threat from right-wing militia and their brethren (see Paul Krugman, "Noonday in the Shade").
47. NRDC (citing *Yale Law Journal,* July 1984), "Gale Norton Biographical Timeline," viewed at http://www.nrdc.org/legislation/norton/app1.asp.
48. Oliver A. Houck, "With Charity for All," *Yale Law Journal,* vol. 93, p. 8, July 1984.
49. NRDC, "Gale Norton Biographical Timeline."
50. Ibid.
51. "Gale Norton's Associations with Anti-Environmental and Wise Use Groups," Clearinghouse on Environmental Advocacy and Research (CLEAR), viewed at http://www.clearproject.org/reports_nortonWU.html.
52. Ibid.

53. NRDC, "Gale Norton Biographical Timeline."
54. Susan Threadgill, "Who's Who," *Washington Monthly,* March 2002, viewed at http://www.washingtonmonthly.com/features/2001/0203.whoswho.html.
55. Ibid.
56. See Friends of the Earth website, http://www.foe.org/camps/eco/interior/manson.html.
57. "Mining Lobbyist Nominated for Key Interior Post," Friends of the Earth press release, December 2001, viewed at http://www.foe.org/camps/eco/interior/watsonpr.html.
58. Susan Threadgill, "Who's Who."
59. "George W. Bush's Anti-Environmental Advisors: Beyond the Official Bios," Environmental Working Group, viewed at http://www.ewg.org/reports/GeorgeWBush/enviro_advisors.html.
60. Phillip Babich, "Shafted," salon.com, December 11, 2003, viewed at http://www.salon.com/tech/feature/2003/12/11/griles.
61. "In the Chief's Corner," Public Employees for Environmental Responsibility, viewed at http://www.peer.org/Chiefchambers/ChiefChamberspage.html.
62. David Halvarg, "Gale Norton, Spencer Abraham, Don Evans: A Team Only an Oil Man Could Love," *The Nation,* January 29, 2001.

3| The First Round

1. "Why Is Election 2000 So Close?" CNN *Crossfire,* viewed at http://www.cnn.com/TRANSCRIPTS/0010/26/cf.00.html.
2. "Oppose the Kyoto Protocol Because It Is Ineffective, Inadequate and Unfair to America," GeorgeWBush.com, viewed at http://web.archive.org/web/20010413031333/www.georgewbush.com/issues/environment.html.
3. "Energy," GeorgeWBush.com, viewed at http://web.archive.org/web/20010204002000/www.georgewbush.com/issues/energy.html.
4. Douglas Jehl, "Bush, in Reversal, Won't Seek Cut in Emissions of Carbon Dioxide," *New York Times,* March 14, 2001.

5. "On the Shoulders of Giants," NASA Earth Observatory Library, viewed at http://www.earthobservatory.nasa.gov/Library/Giants/Arrhenius/arrhenius_2.html.
6. James Hansen, "Global Warming, Playing Dice, and Berenstain Bears," NASA Goddard Institute for Space Study website, January 2000, viewed at http://www.giss.nasa.gov/research/intro/hansen_08/.
7. "World Meteorological Association Statement on the Status of the Global Climate in 2003" (WMO No. 966), viewed at http://www.wmo.ch/web/wcp/wcdmp/home.html.
8. Ibid.
9. Michael Oppenheimer, interview by Robert F. Kennedy, Jr., February 2004.
10. Associated Press, March 4, 2004, viewed at http://msnbc.msn.com/id/4418584/
11. Michael Oppenheimer, interview by Robert F. Kennedy, Jr., February 2004.
12. Michael Peschardt, "Asia-Pacific Climate Change Killing Coral Reefs," BBC News, August 17, 1999, viewed at http://news.bbc.co.uk/1/hi/world/asia-pacific/422759.stm.
13. Michael Oppenheimer, interview by Robert F. Kennedy, Jr., February 2004.
14. Peter Schwartz and Doug Randall, "An Abrupt Climate Change Scenario and Its Implications for United States National Security," Environmental Defense, October 2003, viewed at http://www.ems.org/climate/pentagon_climate_change.html.
15. Ibid.
16. "FAQ: Global Climate Change," United Nations Framework Convention on Climate Change, viewed at http://unfccc.int/press/dossiers/factsheet.html.
17. Michael Ball, "EPA: Whitman Coasts Through Nomination Vote," *Greenwire,* January 31, 2001.
18. Ron Suskind, *The Price of Loyalty* (New York: Simon & Schuster, 2004), p. 101.
19. Ibid.
20. Paul O'Neill, former U.S. Secretary of the Treasury, interview by Robert F. Kennedy, Jr., January 2004.

21. Paul O'Neill, "Science, Politics and Global Climate Change," (Pittsburgh, Pa.: Alcoa, 1998), p. 5.
22. Suskind, *The Price of Loyalty,* p. 114.
23. Ibid., p 102.
24. Ibid., p. 120.
25. Ibid., pp. 121-22.
26. Ibid., p. 122.
27. Ibid.
28. Gregg Easterbrook, "Hostile Environment," *New York Times,* August 19, 2001.
29. Katharine Q. Seelye, "Often Isolated, Whitman Quits as E.P.A. Chief," *New York Times,* May 22, 2003.
30. "NRDC Denounces EPA's Proposal to Withdraw New Arsenic-in-Tap-Water Standard; Group Says Move Is Unwarranted and Illegal—and Vows to Sue," NRDC press release, March 20, 2001, viewed at http://www.nrdc.org/media/pressreleases/010320.asp.
31. Douglas Jehl, "E.P.A. Abandons New Arsenic Limits for Water Supply," *New York Times,* March 21, 2001, p. A1.
32. Physicians for Social Responsibility, "Clinton Era Arsenic Standard Upheld," *Environment and Health Update,* October/November 2001, viewed at http://www.psr.org/documents/psr_doc_0/program_3/eupdate101101.pdf; "Bush Administration, Ignoring National Academy Of Sciences, Sticks With Clinton Arsenic-In-Tap-Water Standard," NRDC, October 31, 2001, viewed at http://www.nrdc.org/media/pressreleases/011031.asp.
33. "House Rebukes Bush on Arsenic Standards," Comtex, July 30, 2001, viewed at http://www.enn.com/news/wire-stories/2001/07/07302001/arsenic_44465.asp.
34. Erik Olson, interview by Robert F. Kennedy, Jr., February 2004.
35. David Frum, *The Right Man: The Surprise Presidency of George W. Bush* (New York: Random House, 2003), p. 7.
36. Gregg Easterbrook, "Hostile Environment," *New York Times Magazine,* August 19, 2001, section 6, p. 40.
37. Douglas Jehl, "EPA Delays Its Decision on Arsenic," *New York Times,* April 19, 2001, p. A14.

38. "Arsenic in Drink Water, 2001 Update," National Academy of Science, viewed at http://www.nap.edu/books/0309076293/html.
39. Eric Pianin, "EOA to Urge Tighter Rules for Arsenic; Report Raises Agency Concern About Drinking Water Limits," *Washington Post,* September 11, 2001, p. A1.
40. Katherine Q. Seelye, "Arsenic Standard for Water Is Too Lax, Study Concludes," *New York Times,* September 11, 2001, p. A20.
41. "EPA Announces Arsenic Standard for Drinking Water of 10 Parts per Billion," EPA Newsroom, November 1, 2001, viewed at http://www.epa.gov/epahome/headline_110101.htm.
42. Erik Olson, interview, February 2004.
43. "Arsenic in Drinking Water," U.S. Environmental Protection Agency, viewed at http://www.epa.gov/safewater/arsenic.html.
44. Cyril T. Zaneski, "Rule Breakers," GovExec.com, January 1, 2002, viewed at http://www.govexec.com/features/0102/0102s4.htm.

4| Cost-Benefit Paralysis

1. "OIRA Q&A's," Office of Management and Budget, February 25, 2002, viewed at http://www.whitehouse.gov/omb/inforeg/qa_2-25-02.pdf.
2. Mary Anne Calamas, assistant to John Graham, e-mail to David Ludlum, April 26, 2004.
3. Judy Dagastino, Advisory Board Coordinator, American Council on Science and Health, interview by David Ludlum, April 15, 2004. Graham served from 1994 to 2001, until he left to work for the government.
4. "Panic Attack: ACSH Fears Nothing but Fear Itself," Center for Media and Democracy, viewed at http://www.prwatch.org/prwissues/1998Q4/panic.html.
5. "Testimony of Professor Lisa Heinzerling Concerning the Nomination of John D. Graham to Be Administrator of the Office of Information and Regulatory Affairs, Office of Manage-

ment and Budget," May 10, 2001, viewed at http://www.citizen.org/congress/regulations/graham/heinzerling_testimony.html.

6. Steve Weinberg, "Mr. Bottom Line," OnEarth Magazine, Spring 2003, viewed at http://www.nrdc.org/onearth/03spr/graham1.asp.
7. "Dr. John Graham's Biography and Research Interests," HCRA Harvard, viewed at http://www.hcra.harvard.edu/nomination.
8. Karl Kelsey, professor at Harvard School of Public Health, interview by Robert F. Kennedy, Jr., March 2004.
9. "HCRA Sources of Support," HCRA Harvard, viewed at http://www.hcra.harvard.edu/unrestricted.html.
10. "Advisory Council," HCRA Harvard, viewed at http://www.hcra.harvard.edu/advisory.html.
11. "Executive Council," HCRA Harvard, viewed at http://www.hcra.harvard.edu/executive.html.
12. "National Organizations Make the Case Against John Graham to the Senate Governmental Affairs Committee," Public Citizen, viewed at http://www.citizen.org/congress/regulations/graham/govtaffairs.html.
13. Amy Charlene, "Harvard's John Graham Releases Results of Cost-Benefit Analysis of Air Bag Safety," Risk World, March 25, 1997, viewed at http://www.riskworld.com/NEWS/97q1/nw7aa028.htm.
14. "Study Shows Airbags a Worthwhile Investment: Risk to Children Must Be Addressed," Harvard School of Public Health press release, November 4, 1997, viewed at http://www.hsph.harvard.edu/press/releases/press11041997.html.
15. Tammy O. Tengs and John D. Graham, "The Opportunity Costs of Haphazard Social Investments in Life Saving," in *Risks, Costs, and Lives Saved: Getting Better Results from Regulation,* Robert W. Hahn, ed. (New York: Oxford University Press, 1996).
16. "John Graham's Distortions: Study Was Used by Him to Mislead Congress and the Media About the Cost of Health, Safety and Environmental Regulation," Public Citizen, August 28, 2001, viewed at http://www.citizen.org/congress/regulations/bush_admin/articles.cfm?ID=5002.

17. Ibid.
18. "Testimony of Professor Lisa Heinzerling Concerning the Nomination of John D. Graham."
19. "John Graham's Distortions."
20. "Industry Ally, Hostile to Health and Environmental Safeguards, Poor Choice for Key Regulatory Post in Bush Administration," Public Citizen, viewed at http://www.citizen.org/congress/regu lations/graham/edit.html.
21. Letter from 53 Academics to the Senate Governmental Affairs Committee, May 9, 2001, viewed at http://www.citizen.org/con gress/regulations/graham/academics.html.
22. "Senate Confirms Enemy of Regulations for Top Regulatory Post," NRDC press release, July 20, 2001, viewed at http:// www.nrdc.org/media/pressReleases/010720.asp.
23. "Rulemaking: OMB's Role in Reviews of Agencies' Draft Rules and the Transparency of Those Reviews," U.S. General Accounting Office, September 2003 (GAO 03-929), viewed at http://www.gao.gov/new.items/d03929.pdf.
24. "Making Sense of Regulation: 2001 Report to Congress on the Cost and Benefits of Regulations and Unfunded Mandates on State, Local and Tribal Entities," Office of Management and Budget, December 2001, viewed at http://www.whitehouse. gov/omb/inforeg/costbenefitreport.pdf.
25. "Economic Issues: Dr. Weida's Reports," Grace Factory Farm Project, viewed at http://www.factoryfarm.org/topics/economic/ weida/#manure_management_analysis.
26. "Testimony of Richard J. Dove," Senate Committee on Government Affairs, March 13, 2002, viewed at http://www.senate. gov/~gov_affairs/031302dove.htm.
27. See "Waste: Facts and Data," Grace Factory Farm Project, viewed at http://www.factoryfarm.org/topics/waste/facts/.
28. "Testimony of Richard J. Dove."
29. Arthur Hirsch, "Meet the Choptank's New Menace: Deadly, Shape-Changing, Tough Algae," *Baltimore Sun,* March 16, 1993.
30. Lynn Grattan et al., "Learning and memory difficulties after environmental exposure to waterways containing toxin-pro-

ducing Pfiesteria or Pfiesteria-like dinoflagellates," *The Lancet,* vol. 352 (1998), pp. 530-39.

31. "Cattle-Theft Case Won't Go to Trial As South Dakota Man Has Reached a Plea Agreement in Custer County," *Omaha World-Herald,* August 22, 2002.
32. "Rulemaking: OMB's Role in Reviews of Agencies' Draft Rules and the Transparency of Those Reviews," U.S. General Accounting Office, September 2003 (GAO 03-929), pp. 103-5, viewed at http://www.gao.gov/new.items/d03929.pdf.
33. "Statement of the American Farm Bureau Federation to the Clean Air, Wetlands, Private Property and Nuclear Safety Subcommittee, Senate Environment and Public Works Committee Regarding Air Quality Standards," presented by Adam Sharp, Assistant Director Governmental Relations, American Farm Bureau Federation, October 22, 1997, viewed at http://epw.senate.gov/105th/sharp.htm.

5| Science Fiction

1. See the Luntz memo, "The Environment: A Cleaner, Healthier, Safer America," Environmental Working Group website, viewed at http://www.ewg.org/briefings/luntzmemo/pdf/Luntz Research_environment.pdf.
2. "Bush Proposes to Cut Nondefense R&D Over the Next Five Years to Reduce Deficit," American Association for the Advancement of Science, May 7, 2004, viewed at http://www.aaas.org/spp/rd/proj05p.pdf.
3. Michael Oppenheimer, interview by Robert F. Kennedy, Jr., February 2004.
4. "EPA's Response to the World Trade Center Collapse: Challenges, Successes, and Areas for Improvement," EPA Office of Inspector General, August 21, 2003, viewed at http://www.epa.gov/oigearth/reports/2003/WTC_report_20030821.pdf.
5. "GAO Confirms EPA Stripped Ombudsman Authority," Rep. Jerrold Nadler press release, November 14, 2002, viewed at http://www.house.gov/nadler/archive107/Ombudsman_111402.htm.

6. Mike Soraghan, "EPA Leader's Rulings on Citigroup Probed," *Denver Post,* March 11, 2002.
7. Hugh Kaufman, interview by Robert F. Kennedy, Jr., January 2004.
8. Ibid.
9. Juan Gonzalez, "It's Public be Damned at the EPA," *New York Daily News,* viewed at http://www.nydailynews.com/front/story/112247p-101268c.html.
10. "Ombudsman Sues to Prevent EPA From Dissolving His Office," *Greenwire,* January 11, 2002.
11. "Ombudsman Quits in Protest After Agency Transfers His Office," *Greenwire,* April 23, 2002.
12. "OSHA Panel Orders Agency to Reinstate Kaufman to Ombudsman's Office," *Greenwire,* July 16, 2002.
13. "EPA's Response to the World Trade Center Collapse," pp. 7, 11.
14. Ibid., p. 17.
15. Ibid, p. 8.
16. Jennifer Lee, "Details Emerge on Post 9-11 Clash Between White House and E.P.A.," *New York Times,* October 10, 2003.
17. "Trade Center Debris Pile Was a Chemical Factory, Says New Study," UC Davis, September 10, 2003, viewed at http://delta.ucdavis.edu/wtc_debris.htm.
18. "Analysis Reveals Over 50% of Responders Experience Pulmonary, ENT and/or Mental Health Symptoms One Year Following NY Terrorist Attacks," Mount Sinai Medical Center press release, January 27, 2003, viewed at http://www.wtcexams.org/pressrelease-20030127.html.
19. Dan Tishman, president of Tishman Construction, interview by Robert F. Kennedy, Jr., January 2004.
20. Hugh Kaufman, interview, January 2004.
21. Peter Waldman, "EPA Bans Staff from Discussing Issue of Perchlorate Pollution," *Wall Street Journal,* April 28, 2003, p. A3.
22. John J. Fialka, "Mercury Threat to Kids Rising, Unreleased EPA Report Says," *Wall Street Journal,* February 20, 2003; see also http://www.nrdc.org/bushrecord/articles/br_1258.asp?t=t.

23. "America's Children at Risk—Mercury in Fish," Environmental Protection Agency, viewed at http://www.epa.gov/envirohealth/children/emerging_issues/fish.htm.
24. "Methylmercury: Epidemiology Update," presentation by Kathryn R. Mahaffey, U.S. EPA, at the National Forum on Contaminants in Fish, San Diego, January 26, 2004, viewed at http://www.epa.gov/waterscience/fish/forum/2004/presentations/monday/mahaffey.pdf.
25. "Federal Advisory For Mercury-Contaminated Fish Proves We Need To Remove Mercury From Commerce, Says NRDC," NRDC, viewed at http://www.nrdc.org/media/pressreleases/040319.asp.
26. Marc Kaufman, "Women, Children Warned About Tuna Consumption," *Washington Post,* March 19, 2004.
27. David Ludlum, interview with Vas Aposhian, May 27, 2004.
28. Jeremy Symons, "How Bush and Co. Obscure the Science," *Washington Post,* July 13, 2003.
29. "Report by E.P.A. Leaves Out Data on Climate Change," *New York Times,* June 19, 2003.
30. Guy Gugliotta and Eric Pianin, "EPA Withholds Air Pollution Analysis," *Washington Post,* July 1, 2003, p. A03, viewed at http://www.washingtonpost.com/ac2/wp-dyn/A54598-2003Jun30?language=printer.
31. Associated Press, "New Climate Plan Draws Heat," July 24, 2003, viewed at http://www.cbsnews.com/stories/2003/06/19/politics/main559380.shtml.
32. Michael Oppenheimer, interview by Robert F. Kennedy, Jr., February 2004.
33. "Research on Oil and Gas Practices," Politics and Science, viewed at http://www.house.gov/reform/min/politicsandscience/example_oil_and_gas.htm.
34. Letter from Rep. Henry A. Waxman to EPA Secretary Christine Todd Whitman, October 8, 2002, viewed at http://www.henrywaxman.house.gov/issues.htm.
35. Jeff Ruch, interview by Robert F. Kennedy, Jr., January 2004.
36. Tom Hamburger, "Water Saga Illuminates Rove's Methods," *Wall Street Journal,* July 30, 2003.

37. "Klamath River Water Levels Set Under Political Pressure," PEER press release, viewed at http://www.peer.org/press/286.html.
38. Mike Kelly, interview by Robert F. Kennedy, Jr., August 29, 2003.
39. Michael Grunwald, "Warnings on Oil Drilling Reversed," *Washington Post,* April 7, 2002.
40. Michael Grunwald, "Interior: Drilling Won't Violate Polar Bear Pact; Stance Contradicts Wildlife Agency's Drafts," *Washington Post,* January 18, 2002.
41. *The Fund for Animals v. Norton,* 294 F. Supp. 2d 92 (D. D.C. 2003), viewed at http://www.dcd.uscourts.gov/02-2367.pdf.
42. Roger Kennedy, interview by David Ludlum, December 2003.
43. Perry Beeman, "Ag scientists feel the heat," *Des Moines Register,* December 1, 2002, viewed at http://www.dmregister.com/business/stories/c4789013/19874144.html.
44. James Zahn, former Agricultural Research Service microbiologist, interview by Robert Kennedy, Jr., April 2002.
45. "Battle Over IPCC Chair Renews Debate on U.S. Climate Policy," *Science,* April 12, 2002.
46. Allyn Sappa, interview by David Ludlum, January 2004.
47. John Paul, interview by David Ludlum, January 2004.
48. David Michaels et al., "Advice Without Dissent," *Science,* October 25, 2002.
49. Deposition of William Banner Jr., M.D., June 13, 2002, in *Rhode Island v. Lead Industries Assoc, Ind* (Sup Ct. R.I.) (No.995226), p. 133; cited in "Turning Lead Into Gold: How the Bush Administration is Poisoning the Lead Advisory Committee at the CDC," by U.S. Rep. Edward J. Markey, October 8, 2002, viewed at http://www.mindfully.org/Health/2002/Lead-Into-Gold-MARKEY8oct02.htm.
50. Bruce Lanphear, interview by David Ludlum, January 2004.
51. Tom Devine, interview by David Ludlum, January 2004.
52. Wes Wilson, interview by David Ludlum, January 2004.
53. "Park Service Set to Outsource Staff," PEER press release, January 27, 2003, viewed at http://www.peer.org/PARK_SERVICE/press/311.html.

54. Wes Wilson, interview, January 2004.
55. Tyrone B. Hayes et al., "Hermaphroditic, demasculinized frogs after exposure to the herbicide atrazine at low ecologically relevant doses," *PNAS* 99:5476-5480 (2002); viewed at http://www.pnas.org./cgi/content/abstract/99/8/6576 (abstract).
56. Shanna H. Swan et al., "The Question of Declining Sperm Density Revisited: An Analysis of 101 Studies Published 1934-1996," *Environmental Health Perspective* 108:961-966 (2000), viewed at http://ehp.niehs.nih.gov/members/2000/108p961-966swan/swan-full.html.
57. Emily Green, "Regulators to Let Maker Test Chemical Levels," *Los Angeles Times,* November 1, 2003.
58. "Wetlands Pollute, Says Study OKed by EPA: EPA Biologist Resigns in Protest; Study Clears Way for SW Florida Developments," PEER, October 22, 2003, viewed at http://www.peer.org/press/403.html.
59. Holly Greening, Tampa Bay National Estuary Program, interview by David Ludlum, January 17, 2004.
60. Bruce Boler, interview by David Ludlum, February 17, 2004.
61. Federal Register, vol. 68 (2003); p. 54,023 Rick Weiss, "Peer Review Plan Draws Criticism," *Washington Post,* January 15, 2004.
62. Karl Kelsey, interview by Robert F. Kennedy, Jr., March 2004.
63. "Scientific Integrity in Policy Making," Union of Concerned Scientists, February 2004, viewed at http://www.ucsusa.org/publications/report.cfm?publicationID=730.
64. Michael Oppenheimer, interview by Robert F. Kennedy, Jr., February 2004.

6| Blueprint for Plunder

1. "Energy/Natural Resources: Long-Term Contribution Trends," Open Secrets, viewed at http://www.opensecrets.org/industries/indus.asp?Ind=E.
2. "Condoleezza Rice Profile," Open Secrets, viewed at http://www.opensecrets.org/bush/cabinet/cabinet.rice.asp; see also, "The

Bush-Cheney Energy Plan: Players, Profits and Paybacks," NRDC, viewed at http://www.nrdc.org/air/energy/aplayers.asp.

3. "The Bush-Cheney Energy Plan."
4. "The Cheney Energy Task Force," NRDC, viewed at http://www.nrdc.org/air/energy/taskforce/tfinx.asp; see also, "The Bush-Cheney Energy Plan."
5. See http://www.archives.gov/federal_register/public_laws/acts.html; see also John W. Dean, "GAO v. Cheney Is Big Time Stalling," at http://writ.news.findlaw.com/dean/20020201.html.
6. "The Bush-Cheney Energy Plan."
7. "Company with Energy Ties to Cheney's Energy Task Force Faces Criminal Indictment for Gaming California's Electricity Market," Online Journal, viewed at http://www.onlinejournal.com/Special_Reports/032604Leopold/032604leopold.html.
8. Ibid.; see also Carl Pope and Paul Rauber, "The Bush Administration: Bright Light Must Shine on Energy Policymaking," *Los Angeles Times,* December 21, 2003, p. M2.
9. "The New Administration: Contributors to Gale Norton's '96 Senate Run," Center for Responsive Politics, viewed at http://www.opensecrets.org/alerts/v6/alertv6_05.asp.
10. "The Bush-Cheney Energy Plan."
11. Ibid.
12. "ExxonMobil Corporation Announces Estimated First Quarter 2001 Results," ExxonMobil press release, April 23, 2001, viewed at http://www2.exxonmobil.com/Corporate/Newsroom/NewsReleases/NewsReleaseIndex.asp.
13. "Slower, Costlier and Dirtier: A Critique of the Bush Energy Plan," NRDC, viewed at http://www.nrdc.org/air/energy/scd/scdinx.asp.
14. "Letter From David M. Walker, Comptroller General of the United States to Rep. Henry Waxman," January 30, 2002. viewed at http://www.house.gov/reform/min/pdfs/pdf_inves/pdf_energy_cheney_jan_30_gao_let.pdf. This suit was dismissed and the GAO did not appeal the dismissal. See Judicial Watch, "GAO Drops Lawsuit Against Cheney Energy Lawsuit," at http://www.judicialwatch.org/3140.shtml.

15. "GAO Suit Against Cheney Energy Task Force Rejected," NRDC, viewed at http://www.nrdc.org/bushrecord/articles/br_1199.asp?t=t.
16. "GAO Halts Lawsuit Over Cheney Energy Files," NRDC, viewed at http://www.nrdc.org/bushrecord/articles/br_1253.asp?t=t.
17. "Supreme Court Will Review Lawsuit Over Cheney Energy Meetings," CNN.com, viewed at http://www.cnn.com/2003/ALLPOLITICS/12/15/scotus.cheney.ap/.
18. Ibid.
19. "High Court Justice a Cheney Guest," CBSNews.com, viewed at http://www.cbsnews.com/stories/2004/02/11/politics/main599686.shtml.
20. "Justice's Blind," *Salt Lake Tribune,* February 12, 2004, p. A14.
21. "The Cheney Energy Task Force: How NRDC Brought the Records to Light," NRDC, viewed at http://www.nrdc.org/air/energy/taskforce/bkgrd.asp.
22. "Biography of Richard Cheney," InfoPlease, viewed at http://www.infoplease.com/ipa/A0882164.html.
23. Paul O'Neill, former U.S. Treasury Secretary, interview by Robert Kennedy, Jr., January 2004.
24. "Energy Task Force: Process Used to Develop the National Energy Policy," General Accounting Office, August 2003, viewed at http://www.mindfully.org/Energy/2003/Energy-Task-Force-GAO25aug03.htm.
25. "The Bush-Cheney Energy Plan."
26. Ibid.
27. Michael Grunwald, "Trade Groups in Lock Step Behind Bush Energy Policy," *Washington Post,* May 30, 2001.
28. "The Bush-Cheney Energy Plan."
29. "Industry Had Extensive Access to Industry Task Force," NRDC, viewed at http://www.nrdc.org/air/energy/taskforce/tfinx.asp.
30. Ibid.
31. Jason Leopold, "Another Slap on the Wrist: How Reliant Energy Withheld Power From California Consumers," *CounterPunch,* February 3, 2003, viewed at http://www.counterpunch.org/leopold02032003.html.

32. Gray Davis, California governor, interview by Robert Kennedy, Jr., January 2004.
33. Jim Hightower, "Tom Delay Thinks He's God's Man in Congress," ZNet, December 17, 2003, viewed at http://www.zmag.org/content/showarticle.cfm?SectionID=43&ItemID=4703; see also Public Citizen website, April 2004, at http://www.citizen.org/pressroom/release.cfm?ID=1686.
34. Mike Allen and Dana Milbank, "Cheney's Role Offers Strengths And Liabilities," *Washington Post,* May 17, 2001.
35. Sandra Sobieraj, "Cheney: Energy Woes Take Time to Fix," Associated Press, May 15, 2001, viewed at http://www.evote.com/index.asp?Page=/news_section/2001-05/05142001Energy.asp.
36. See the following NRDC releases: "Heavily Censored Papers Show Industry Writes Energy Report," at http://www.nrdc.org/media/pressreleases/020327.asp; "Data Shows Industry Had Extensive Access to Cheney's Energy Task Force," at http://www.nrdc.org/air/energy/taskforce/tfinx.asp; "Energy Department Documents Verify Industry Influence," at http://www.nrdc.org/media/pressreleases/020521b.asp.
37. "Data Shows Industry Had Extensive Access to Cheney's Energy Task Force."
38. "The Bush-Cheney Energy Plan."
39. Eric Pianin and Dan Morgan, "Oil Executives Lobbied on Drilling: Two Went to Cheney Task Force to Push for Gulf of Mexico Sale," *Washington Post,* February 27, 2002, p. A1.
40. "Data Shows Industry Had Extensive Access to Cheney's Energy Task Force."
41. "Contributors and Paybacks: The Electric Utility Industry Campaign to Undermine Clean Air Lawsuits to Reduce Pollution," Public Citizen, October 27, 2003, viewed at http://www.whitehouseforsale.org/documents/nsr_fact_sheet.pdf.
42. "Paying to Pollute: Campaign Contributions and Lobbying Expenditures by Polluters Working to Weaken Environmental Laws," US PIRG, April 2004, viewed at http://uspirg.org/reports/payingtopollute2004.pdf.
43. United States Department of Justice, Office of Legal Policy, "New Source Review: An Analysis of the Consistency of En-

forcement Actions With the Clean Air Act and Implementing Regulations," January 2002, Appendix III, viewed at http://www.usdoj.gov/olp/nsrreport.pdf.

44. "Data Shows Industry Had Extensive Access to Cheney's Energy Task Force."
45. Judy Pasternak, "Bush's Energy Plan Bares Industry Clout," *Los Angeles Times,* August 26, 2001, p. A2.
46. Michael Isikoff, "A Plot to Foil the Greens," *Newsweek,* June 4, 2001.
47. Eric Schaeffer, former director of the EPA's Office of Regulatory Enforcement, interview by Robert F. Kennedy, Jr., October 2003 and February 2004.
48. Calculations based on PAC and soft-money contributions tracked by the Center for Responsive Politics database, viewed at http://www.opensecrets.org/softmoney/index.asp.
49. "National Energy Policy," National Energy Policy Development Group, May 2001, viewed at http://www.whitehouse.gov/energy/National-Energy-Policy.pdf.
50. Paul O'Neill, former U.S. Treasury Secretary, interview by Robert Kennedy, Jr., January 2004.
51. Jim VandeHei, "Cheney Sees Little Role for Conservation In Energy Plan Aimed at Boosting Supply," *Wall Street Journal,* May 1, 2001, p. A12.
52. "The Bush-Cheney Energy Plan: How It Fares in the 21st Century," Energy Foundation, viewed at http://www.ef.org/national/NationalAnalysis.pdf.
53. See "America, Oil and National Security: What Government and Industry Data Really Show," National Environmental Trust 2001, viewed at http://www.ef.org/documents/AmericaandOil.pdf; see also "Drilling in Detroit: Tapping Automaker Ingenuity to Build Safe and Efficient Automobiles," Union of Concerned Scientists and Center for Auto Safety, June 2001, viewed at http://www.ucsusa.org/documents/drill_detroit.pdf.
54. Amory Lovins, "Energy Security Facts: Details and Documentation," Rocky Mountain Institute, June 2003, viewed at http://www.rmi.org/images/other/S-USESFbooklet.pdf.
55. Richard Byrne, "Life in the Slow Lane: Tracking Decades of

Automaker Roadblocks to Fuel Economy," Union of Concerned Scientists, July 2003, viewed at http://www.ucsusa.org/clean_vehicles/cars_and_suvs/page.cfm?pageID=1230.
56. Amory Lovins, "Energy Security Facts."
57. "Dangerous Addiction: Ending America's Oil Dependence," NRDC, January 2002, viewed at http://www.nrdc.org/air/transportation/oilsecurity/security.pdf.
58. Jerry Taylor, "Bush's Energy Babble," CATO Institute, September 30, 2000, viewed at http://www.cato.org/dailys/09-30-00.html.
59. "New Campaign Slams Detroit, Washington as U.S. Gas Mileage Hits 22-Year Low," NRDC, May 7, 2003, viewed at http://www.nrdc.org/media/pressreleases/030507.asp.
60. "The Bush Administration's Fuel Cell Fake-Out: White House, Automakers Tout Tomorrow's Oil Security Solution to Deflect and Delay Action Today," NRDC, May 8, 2003, viewed at http://www.nrdc.org/air/transportation/ffuelcell.asp.
61. See "Bush Week in Review, May 7-13," Literal Politics, viewed at http://www.literalpolitics.com/BushWeek/bushweekmay.htm.
62. "Remarks by Secretary of Energy Spencer Abraham, 13th Annual Energy Efficiency Forum, June 12, 2002," viewed at http://www.energy.gov/engine/content.do?PUBLIC_ID=13407&BT_CODE=PR_SPEECHES&TT_CODE=PRESSSPEECH.
63. Tom Kenworthy, "New Mining Rules Reverse Provisions," *USA Today,* October 25, 2001.
64. "President Announces Clear Skies & Global Climate Change Initiatives," press release, February 14, 2002, available at http://www.whitehouse.gov/news/releases/2002/02/20020214-5.html.
65. "Power to Kill: Death and Disease From Power Plants Charged with Violating the Clean Air Act," Clean Air Task Force, July 2001, viewed at http://cta.policy.net/relatives/18300.pdf; see also "Air Quality Management in the United States," National Research Council of the National Academies, January 2004, viewed at http://cta.policy.net/relatives/18300.pdf.
66. Christopher Shays, U.S. Congressman (R-Conn.), interview by Robert F. Kennedy, Jr., September 2003.
67. "Taxpayers Should Finance Cleanups, Bush Says," *Greenwire,* February 25, 2002.

68. "The Truth About Toxic Waste Cleanups: How EPA Is Misleading the Public About the Superfund Program," Sierra Club and US PIRG Education Fund, February 2004, viewed at http://www.uspirg.org/reports/TruthaboutToxicWasteCleanup04.pdf.
69. Ibid.
70. Ibid.
71. Ibid.
72. Paul O'Neill, former U.S. Treasury Secretary, interview by Robert Kennedy, Jr., January 2004.

7| King Coal

1. "Mid-Atlantic Mountaintop Mining Draft Environmental Impact Statement," U.S. EPA, May 2003, viewed at http://www.epa.gov/region3/mtntop/eis.htm.
2. "Mountaintop Mining: Permissive Permitting Doesn't Jibe With Study," *Lexington Herald Leader,* July 22, 2003.
3. Judy Bonds, director of Coal River Mountain Watch, interview by Robert F. Kennedy, Jr., February 2004.
4. "Comments of the Ohio Valley Environmental Coalition (OVEC) on the Draft Programmatic Environmental Impact Statement on Mountaintop Removal Mining/Valley Fill Activities in Appalachia," viewed at http://www.ohvec.org/issues/mountaintop_removal/articles/EIS_social_cultural.pdf.
5. "West Virginia Labor Force Estimates," West Virginia Bureau of Employment Programs, viewed at http://www.state.wv.us/bep/lmi/datarel/DRLMI134.HTM.
6. Judy Bonds, director of Coal River Mountain Watch, interview by Robert F. Kennedy, Jr., February 2004.
7. See "Read My Lips: A Look at Coal Mining Industry Contributions to Bush and the GOP," Open Secrets, March 14, 2001, viewed at http://www.opensecrets.org/alerts/v6/alertv6_11.asp.
8. Tom Hamburger, "A Coal-Fired Crusade Helped Bring Crucial Victory to Candidate Bush," *Wall Street Journal,* June 13, 2001.
9. "100 Days of Bush—Top 10 Paybacks to the Energy Industry,"

Democratic National Committee, viewed at http://www.healthyrecovery.net/Asinine/100days/energy.html.

10. "Reliable, Affordable, and Environmentally Sound Energy for America's Future," National Energy Policy Development Group, 2001, p. xiv, viewed at http://www.whitehouse.gov/energy.
11. Judy Pasternak, "Bush's Energy Plan Bares Industry Clout," *Los Angeles Times,* August 26, 2001.
12. Peter Carlson, "Green for Greenbacks? Bush's Record Under Fire," *Washington Post,* September 9, 2003.
13. Quin Shea, remarks made at the Western Coal Transportation Association Conference, Sante Fe, New Mexico, April 16-18, 2001.
14. Michelle Nijhuis, "Coal Miner's Slaughter: West Virginia Activist Julia Bonds Takes on Mountaintop-Removal Mining," *Grist,* April 14, 2003.
15. Peter Slavin, "Razing Appalachia," *Citizens Coal Council,* May 5, 2002.
16. Jack Spadaro, interview by Robert F. Kennedy, Jr., February 2004.
17. See U.S. Department of Labor website at http://www.dol.gov/_sec/aboutosec/chao.htm.
18. Jack Spadaro, interview by Robert F. Kennedy, Jr., February 2004.
19. "Special Counsel Begins Whistleblower Investigation," *Greenwire,* February 24, 2004.
20. "MSHA Demotes, Moves Ky. Coal-Slurry Spill Whistleblower," *Greenwire,* February 26, 2004.
21. "White House Watch Administration Profiles, Jeffrey Holmstead—Assistant Administrator, Air and Radiation," Earth Justice, viewed at http://www.earthjustice.org/policy/profiles/display.html?ID=1011.
22. "Corporate Shill Enterprise: A Public Citizen Report on Citizens for a Sound Economy, a Corporate Lobbying Front Group," Public Citizen, viewed at http://www.citizen.org/congress/civjus/tort/industry/articles.cfm?ID=798.
23. Frank O'Donnell, interview by David Ludlum, January 29, 2004.

24. "Benchmarking Air Emissions 2001," NRDC, viewed at http://www.nrdc.org/air/pollution/benchmarking/default.asp.
25. Seth Borenstein, "EPA Brass Departing in Uproar," Knight Ridder News Service, September 4, 2003.
26. "Senator Edwards Calls for Resignation of EPA Official," John Edwards press release, July 14, 2003, viewed at http://edwards.senate.gov/press/2003/0713b-pr.html.
27. General Accounting Office, "New Source Review Revisions Could Affect Utility Enforcement Cases and Public Access to Emissions Data," October 2003.
28. U.S. Department of Justice, Office of Legal Policy, "New Source Review: An Analysis of the Consistency of Enforcement Actions With the Clean Air Act and Implementing Regulations," January 2002, Appendix III.
29. "Bush Appointees Gut Air Quality Rule and Give Congress False Information About the Consequences," Public Citizen press release, October 10, 2003.
30. "Letter From Senators Leahy, Jeffords and Lieberman to Inspector General Tinsley About New Source Review Enforcement," October 22, 2003, viewed at http://leahy.senate.gov/press/200310/102203d.html.
31. Senator Patrick Leahy, "Clearing the Air: New Source Review Policy, Regulations and Enforcement Activities," statement for Joint Hearing of the Senate Judiciary Committee and Committee on the Environment and Public Works, July 16, 2002, viewed at http://leahy.senate.gov/press/200207/071602.html.
32. Clean Air Task Force, "Power to Kill: Death and Disease from Power Plants Charged With Violating the Clean Air Act," July 2001, viewed at http://cta.policy.net/relatives/18300.pdf. (EPA consultant Abt Associates contributed the methodology and analysis.)
33. "Senator Edwards Calls for Resignation of EPA Official."
34. See "ToxFAQs for Mercury," Agency for Toxic Substances and Disease Registry, April 1999, at http://www.atsdr.cdc.gov/tfacts46.html.
35. "Twice as Many Newborns Are at Risk for Developmental,

Learning Problems—EPA," *Greenwire,* February 5, 2004, viewed at http://www.ewg.org/news/story.php?id=2234.

36. Dr. David Carpenter, interview by Robert F. Kennedy, Jr., March-April 2004.
37. "Update: National Listing of Fish and Wildlife Advisories," Environmental Protection Agency, EPA-823-F-03-003, May 2003, viewed at http://www.epa.gov/waterscience/fish/advisories/factsheet.pdf.
38. "EPA's Mercury Proposal: More Toxic Pollution for a Longer Time," NRDC, December 5, 2003, available at http://www.nrdc.org/media/pressreleases/031205.asp.
39. EPA to Regulate Mercury and Other Air Toxics Emissions from Coal- and Oil-Fired Power Plants," U.S. EPA fact sheet, viewed at http://www.epa.gov/ttn/atw/combust/utiltox/hgfs1212.html.
40. Quin Shea, remarks made at the Western Coal Transportation Association Conference.
41. Sharon Theimer, "Business Leaders Pay to Wine and Dine Bush Environmental Policymakers," Associated Press, January 7, 2004, viewed at http://www.enn.com/news/2004-01-07/s_11786.asp.
42. "Survey Says: Western Leaders Name Energy Bill, Endangered Species Act Reform as Top Priorities for Congress in 2004," Western Business Roundtable press release, January 27, 2004, viewed at http://www.westernroundtable.com/PR_2004_topten_survey.htm.
43. "Biography of Jim Sims," Western Business Roundtable, viewed at http://www.westernroundtable.com/company/bio_jsims.asp.
44. Michael Shnayerson, "Sale of the Wild," *Vanity Fair,* September 2003.
45. *Virginia Surface Mining and Reclamation Association v. Andrus,* 483 F.Supp. 425 (D.C. Va. 1980).
46. Philip Babich, "Shafted," salon.com, December 11, 2003.
47. Dale Russakoff, "The Unforcer," *Washington Post,* June 6, 1982, p. A1.
48. Ibid.

49. "J. Steven Griles: Coal Lobbyist Nominated for Interior Second in Command," Clearinghouse on Environmental Advocacy and Research, June 2001, viewed at http://www.clearproject.org/reports_griles.html.
50. Ibid.
51. Ibid.
52. Steve Marcy, "Interior Nominee Faces Criticism on Handling of Strip Mine Agency (J. Steven Griles, Office of Surface Mining Reclamation and Enforcement)," *Oil Daily,* December 5, 1985.
53. See *San Diego Union Tribune,* April 11, 1989, cited by Clearinghouse on Environmental Advocacy and Research, viewed at http://www.clearproject.org/reports_griles.html.
54. Mike Soraghan, "Mining Industry Lobbyist Is Norton's Pick for Deputy," *Denver Post,* March 9, 2001, p. A6.
55. Eric Pianin, "Official's Lobbying Ties Decried; Interior's Griles Defends Meetings as Social, Informational," *Washington Post,* September 25, 2002, p. A1.
56. Deputy Secretary J. Steven Griles, "Statement of Disqualification from Matter Involving his Former Employers and Clients," August 1, 2001, viewed at http://www.foe.org/camps/eco/interior/statement1.pdf.
57. Katharine Q. Seelye, "Interior Department Investigates Official's Role as Oil Lobbyist," Corpwatch.org, May 12, 2003, page A27, viewed at http://www.corpwatch.org/article.php?id=6789.
58. Ibid.
59. Friends of the Earth, "The Case Against J. Steven Griles, Deputy Secretary, U.S. Department of the Interior—Background on Coal Bed Methane and the Powder River Basin" viewed at http://www.foe.org/camps/eco/interior/grilescase1.html.
60. Ibid.
61. Ibid.
62. Friends of the Earth, "J. Steven Griles by the Numbers," *Economics of the Earth,* June 2, 2003, viewed at http://www.foe.org/camps/eco/interior/grilesbynumbers.pdf.
63. DOI Office of Inspector General, J. Steven Griles Report of In-

vestigation PI-SI-02-0053-I (March 16, 2004), viewed at http://www.foe.org/griles.pdf.

64. Christopher Lee, "IG Probes Interior's Record on Ethics Rules," *Washington Post,* May 13, 2003, p. A17.
65. Pete Yost, "Environmentals Seek Special Counsel to Investigate No. 2 Official at Interior," *Associated Press,* June 4, 2003, viewed at http://www.enn.com/news/2003-06-04/s_4752.asp.
66. Ibid.
67. DOI Office of Inspector General, J. Steven Griles Report, p. 138.
68. Ibid.
69. Ibid.
70. Ibid.
71. Rick Weiss, "Report Critical of Interior Official: Inspector General Calls Deputy Secretary's Dealings Troubling but Not Illegal," *Washington Post,* March 17, 2004 p. A23.
72. Ibid.
73. "Deputy Interior Secretary Griles' Statement on the Inspector General's Investigation," U.S. Department of Interior press release, March 16, 2004.
74. Rick Weiss, "Report Critical of Interior Official."
75. "J. Steven Griles Meetings with White House Officials," Friends of the Earth, viewed at http://www.ems.org/interior/griles_meetings.html.
76. DOI Office of Inspector General, J. Steven Griles Report, p. 104.
77. Ken Ward, "Corps' Stance on Valley Fills at Issue," *Charleston Gazette,* December 8, 1998, viewed at http://www.wvgazette.com/static/series/mining/CORP1208.html; David Case, "Just Make It Legal: From 'Waste' to 'Fill,'" TomPaine.com, April 15, 2002, viewed at http://www.tompaine.com/feature2.cfm/ID/5460. (Original publisher.)
78. Joe Lovett, interview by David Ludlum, April 1, 2004.
79. *Bragg v. Robertson,* 72 F. Supp. 2d 642 (S.D. W.Va. 1999).
80. See Ken Ward, "Interior Official Maintains Coal Ties: Former Lobbyist Steven Griles Signed Recusal Agreement," *Charleston Gazette,* September 29, 2002.
81. *Bragg v. Robertson,* 72 F.Supp. 2d 642 (S.D. W.Va. 1999) and subsequent cases.

82. Joe Lovett interview; Marty Coyne, "Interior Undermines Key Study on Mountaintop Removal, Enviros Allege," *Greenwire,* August 9, 2003.
83. Letter from J. Steven Griles, Subject: Mountaintop Mining/Valley Fill Issues, October 5, 2001, viewed at http://earthjustice.org/policy/pdf/Griles_10_5_2001_Memo.pdf.
84. "Paybacks: Polluters, Patrons and Personnel," Earthjustice and Public Campaign, September 2002, viewed at http://www.publiccampaign.org/publications/studies/paybacks/Paybacks.pdf.
85. Ibid.
86. Ken Ward, "Dozen in GOP Urge Bush to Leave Water Rule Alone," *Charleston Gazette,* March 31, 2002; "Protect Streams from Mining Waste," Friends of the Earth, March 23, 2004, viewed at http://ga1.org/alert-description.tcl?alert_id=646956.

8| Killing the Energy Bill

1. Carl Hulse, "Advocates of Arctic Drilling Buoyed as House Passes Bill," *New York Times,* April 12, 2003, p. A8; Mike Ewall, "The Energy Bill: The Environment's Worst Nightmare," Energy Justice Network, January 2004, viewed at http://www.energyjustice.net/energybill/jan2004update.html. The bill passed the House in a 247-175 vote.
2. Carl Hulse, "Energy Bill Gives Way to Old One in the Senate," *New York Times,* August 1, 2003, p. A18.
3. "Energy Bill Watch List: Spotlight on House-Senate Energy Conference in Wake of Electricity Blackout, Scandals in Energy Industry," Public Citizen, viewed at http://www.citizen.org/cmep/energy_enviro_nuclear/electricity/energybill/articles.cfm?ID=10395.
4. Carl Hulse, "Accord Reached by Republicans for Energy Bill," *New York Times,* November 15, 2003, p. A1.
5. Carl Hulse, "Energy Bill Could Upend Lawsuits over Gasoline Additive," *New York Times,* November 16, 2003, p. 24.
6. William Niekirk, "Senate Blocks Sweeping Energy Bill," *Seattle Times,* November 22, 2003, viewed at http://seattle

times.nesource.com/html/nationworld/2001798472_enery220.html.

7. "Senate Gives Up on Energy Bill for 2003," CBN.com, November 25, 2003, viewed at http://www.cbn.com/CBNNews?Wire/031125c.asp.
8. Karen Wayland, Natural Resources Defense Council, interview by Robert F. Kennedy, Jr., January 2004.
9. "Senate Adds Rule to Double Ethanol in Gasoline," *New York Times,* June 6, 2003, p. A20. The ethanol rule was meant to double the use of ethanol in gasoline, providing a boon to corn farmers.
10. Ewall, "The Energy Bill." Daschle "sold out to the ethanol lobby and now supports the bill." See also "Senate Adds Rule to Double Ethanol in Gasoline." The ethanol proposal was introduced by Frist and Daschle.
11. Ewall, "The Energy Bill."
12. Carl Hulse, "Senate Starts Energy Debate Under Threat of Filibuster," *New York Times,* Nov. 20, 2003, p. A28.
13. Senator Evan Bayh (D-Ind.), interview by Robert F. Kennedy, Jr., December 2003.
14. "The Price of Governance," *Wall Street Journal,* November 24, 2004, p. A14.
15. Karen Wayland, Natural Resources Defense Council, interview by Robert F. Kennedy, Jr., January 2004.
16. Senator Evan Bayh (D-Ind), interview by Robert F. Kennedy, Jr., December 2003.
17. Greg Wetstone, Natural Resources Defense Council, interview by Robert F. Kennedy, Jr., January 2004.
18. Ewall, "The Energy Bill."
19. Senator John Sununu (R-N.H.), interview by Robert F. Kennedy, Jr., February 2004.
20. The House passed the bill with a vote of 246-180. Ewall, "The Energy Bill."
21. "Archer-Daschle-Midland," *Wall Street Journal,* November 21, 2003, p. A12
22. Greg Wetstone, Natural Resources Defense Council, interview by Robert F. Kennedy, Jr., January 2004.

23. Ibid. The final vote was 57 to 40, not 58, because Frist changed his vote procedurally in order to preserve his right to bring the legislation up for another vote. See also Carl Hulse, "Senate Blocks Energy Bill; Backers Vow to Try Again," *New York Times,* November 22, 2003, p. A1.
24. Greg Wetstone, Natural Resources Defense Council, interview by Robert F. Kennedy, Jr., January 2004.
25. "MTBE in Drinking Water," Environmental Working Group, viewed at http://www.ewg.org/issues/MTBE/20031001/report.php.
26. "I Don't Want My MTBE: GOP Protects Fuel Additive Manufacturers Despite Drinking Water Contamination," Public Citizen, viewed at http://www.citizen.org/cmep/energy_enviro_nuclear/electricity/energybill/2003/articles.cfm?ID=10546.
27. "MTBE Fate and Transport," New Jersey Department of Environmental Protection, viewed at http://www.state.nj.us/dep/dsr/mtbe/fate_and_transport.htm.
28. "MTBE: What the Oil Companies Knew and When they Knew It," Environmental Working Group, viewed at http://www.ewg.org/reports/withknoweldge/index.php.
29. "Minutes of the MTBE-Environment Meeting," The Hague, October 24-25, 1991.
30. "MTBE: What the Oil Companies Knew."
31. "Occurrence of the Gasoline Additive MTBE in Shallow Ground Water in Urban and Agricultural Areas," U.S. Geological Survey, viewed at http://sd.water.usgs.gov/nawqa/pubs/factsheet/fs114.95/fact.html.
32. "MTBE cleanup in the U.S. will cost at least $29 billion," viewed at http://eces.org/archive/ec/pollution/mtbe.shtml; Matthew Hagemann et al., "Cleanup Cost for MTBE Contamination of Drinking Water Supplies," report forthcoming.
33. "I Don't Want My MTBE: GOP Protects Fuel Additive Manufacturers Despite Drinking Water Contamination."

9| National Security

1. President George W. Bush, "Remarks by the President on the One-Year Anniversary of the U.S. Department of Homeland Security," March 2, 2004, available at http://www.whitehouse.gov/news/releases/2004/03/20040302-2.html.
2. EPA Risk Management Plan (Facility ID: 100000025215) available by request at http://www.rtknet.org/rmp/NJ.php.
3. Richard Pienciak, "Toxic Time Bombs at Chemical Plants: Terror Strike Could Endanger Millions," *New York Daily News,* July 14, 2002.
4. See Margaret Kriz, "In Sight: Chemical Targets," *The Record,* August 10, 2003, p. 1.
5. Rick Hind, Legislative Director of the Greenpeace Toxics Campaign, interview by David Ludlum, May 13, 2004.
6. Rand Beers, former Special Assistant to the President and Senior Director for Combating Terrorism for the National Security Council, interview by Robert F. Kennedy, Jr., February 2004.
7. Margaret Kriz, "Bush Not Doing Enough to Protect Chemical Plants, Critics Contend," *National Journal,* August 7, 2003, viewed at http://www.govexec.com/dailyfed/0803/080703nj3.htm.
8. "Al Qa'ida Chemical, Biological, Radiological, and Nuclear Threat and Basic Countermeasures," Bulletin from the Department of Homeland Security's National Infrastructure Protection Center, no. 03-003, February 12, 2003, available at http://www.nipc.gov/warnings/infobulletins/2003/ib03003.htm.
9. *Congressional Record,* volume 149, pp. S289-S305 (January 14, 2003), viewed at http://www.fas.org/sgp/congress/2003/s157.html.
10. Senator John Corzine (D-N.J.), interview by Robert F. Kennedy, Jr., April 2004.
11. "Chemical Reaction Despite Terrorism Threat, Chemical Industry Succeeds in Blocking Federal Security Regulations," Common Cause, January 27, 2003, viewed at http://www.commoncause.org/publications/jan03/012703_chemical_reaction.pdf.

12. Ibid.
13. Paul Rosenzweig, "The Chemical Security Act: Using Terrorism as an Excuse to Criminalize Productive Economic Activity," Executive Memorandum #833, Heritage Foundation, September 12, 2002, viewed at http://www.heritage.org/Research/HomelandDefense/em833.cfm.
14. See http://www.nationalreview.com/comment/comment-logomasini091702.asp.
15. Amy Ridenour, "When It Comes to Safeguarding Chemical Facilities, the EPA is No Defense Department," *National Policy Analysis,* October 2002, viewed at http://www.nationalcenter.org/NPA436.html.
16. "Despite Terrorism Threat, Chemical Industry Succeeds in Blocking Federal Security Regulations," Common Cause, January 27, 2003, viewed at http://www.commoncause.org/publications/jan03/012703_2.htm.
17. Ledyard King and Beth Gorczyca, "Chemical Plants Tighten Security—Senate Might Require Industries to Assess Possible Terrorist Threats," *Herald-Dispatch,* September 26, 2002.
18. September 10, 2002, letter from Senators Inhofe (Okla.), Arlen Specter (Pa.), Bob Smith (N.H.), Kit Bond (Mo.), George Voinovich (Ohio), Mike Crapo (Idaho), Pete Domenici (N.M.); George Allen letter to Tom Daschle and Trent Lott, August 23, 2002.
19. "Chemical Reaction Despite Terrorism Threat."
20. Margaret Kriz, "In Sight: Chemical Targets."
21. Margaret Kriz, "Bush Not Doing Enough to Protect Chemical Plants, Critics Contend."
22. Ibid.
23. Senator John Corzine (D-N.J.), interview by Robert F. Kennedy, Jr., April 2004.
24. "Homeland Security: Voluntary Initiatives Are Under Way at Chemical Facilities, But the Extent of Security Preparedness Is Unknown," General Accounting Office Policy Papers, March 2003, viewed at http://www.gao.gov/new.items/d03439.pdf.
25. Meghan Purvis and Julia Bauler, "Irresponsible Care: The Failure of the Chemical Industry to Protect the Public from Chem-

ical Accidents," U.S. Public Interest Research Group Education Fund, April 2004, p. 5, viewed at http://uspirg.org/reports/IrresponsibleCare2004.pdf.

26. "Homeland Security: Voluntary Initiatives Are Under Way."
27. Carl Prine, "Lax Security Exposes Lethal Chemical Supplies," *Pittsburgh Tribune Review,* April 7, 2002.
28. "Chemical Plant Security Breaches in the News," Working Group on Community Right-to-Know, viewed at http://www.crtk.org/detail.cfm?docID=26&cat=spills%20and%20emergencies.
29. U.S. Senate Report 107-342, November 15, 2002, viewed at http://www.congress.gov/cgi-bin/cpquery/R?cp107:FLD010:@1(sr342); Tom Ridge, "Testimony before the Senate Committee on Environment and Public Works Concerning Nuclear Power Plant Security," July 10, 2002, available at http://www.nrc.gov/what-we-do/safeguards/response-911.html.
30. *NOW With Bill Moyers,* aired on PBS, March 21, 2003.
31. Ron Gatto, director of Environmental Security for Westchester County, interview by Robert F. Kennedy, Jr., August 2003.
32. See Tarannum Kamlani, "Life in a Reactor's Shadow," MSNBC, June 5, 2002, viewed at http://www.msnbc.msn.com/id/3072034/; see also "Review of Emergency Preparedness of Areas Adjacent to Indian Point and Millstone," James Lee Witt Associates, LLC, January 10, 2003 [Witt Report], viewed at http://www.wittassociates.com/NYReport.zip.
33. Public Citizen, Petition Pursuant to 10 Cfr §2.206, Indian Point Unit 2, Docket No. 50-247, available at http://www.citizen.org/cmep/energy_enviro_nuclear/nuclear_power_plants/decomissioning/articles.cfm?ID=5861.
34. Kyle Rabin, Riverkeeper, e-mail to David Ludlum, May 12, 2004.
35. See V. L. Sailor et al., "Severe Accidents in Spent Fuel Pools in Support of Generic Safety Issue 82," NUREG/CR-4982, Washington, D.C.: NRC, July 1987.
36. "The Mastermind," *60 Minutes II,* March 5, 2003.
37. "Fresh FBI terror warning," BBC News, November 15, 2002, viewed at http://news.bbc.co.uk/2/hi/americas/2481799.stm.
38. "Security Gap: A Hard Look at the Soft Spots in Our Civilian

Nuclear Reactor Security," U.S. Rep. Edward Markey, March 25, 2002.

39. Title 10 of the Code of Federal Regulations, Section 50.13 (10 CFR §50.13).
40. Douglas Pasternak, "A Nuclear Nightmare," *U.S. News & World Report,* September 17, 2001, viewed at http://www.nci.org/01/09/09-3.htm.
41. "Report of Investigation: Entergy Nuclear Northeast, Indian Point #2; Security Services IP2-431," Keith Logan, January 25, 2002.
42. *Riverkeeper, Inc. v. Collins,* 2004 U.S. App. LEXIS 3398 (2d Cir. 2004).
43. "Nuclear Regulatory Commission Oversight of Security at Commercial Nuclear Power Plants Needs to Be Strengthened," U.S. General Accounting Office, Sept. 2003, viewed at http://www.gao.gov/cgi-bin/getrpt?GAO-03-752.
44. Ibid.
45. Ibid.
46. Matthew Wald, "Nuclear Regulatory Agency Lax on Reactor Security, Congressional Audit Finds," *New York Times,* September 28, 2003.
47. "Witt Report," viewed at http://www.wittassociates.com/NY Report.zip.
48. Mary Boyle, "Political Meltdown: Yucca Mountain Supporters Contribute $29.2 Million in Soft Money," Common Cause, March 13, 2002.
49. Ibid.
50. "Corporate Security Spending Only Up Slightly," Conference Board Newsletter, Fall 2003, viewed at http://www.conference-board.org/publications/newsletter/dec2003_security.cfm.

10| What Liberal Media?

1. Senator Gaylord Nelson, founder of Earth Day, at http://www.missouri.edu/~polidjw/Nelson.html; see also Senator Gaylord

Nelson, "All About Earth Day," at http://earthday.wilderness.org/history/.

2. Comments by Gaylord Nelson to Robert F. Kennedy, Jr., April 22, 1990.
3. See "Diminishing Returns: World Fisheries Under Pressure," World Resources Institute, 1998-1999, viewed at http://pubs.wri.org/pubs_content_text.cfm?ContentID=1390.
4. NRDC, "Global Warming Puts the Arctic on Thin Ice," NRDC, viewed at http://www.nrdc.org/globalWarming/qthinice.asp.
5. "Key Facts: Race, Ethnicity & Medical Care," Kaiser Family Foundation, June 2003, viewed at http://www.kff.org/minority health/6069=index.cfm; see also, "Racial Bias Has Effect on Health Stress of Unequal Access, Socioeconomic Status Play Roles, says Expert," *Akron Beacon Journal,* June 15, 2003.
6. "Study Sees Mass Extinctions Via Warming," MSNBC, January 8, 2004, viewed at http://msnbc.msn.com/id/3897120.
7. "Mixed Media," *Sierra Club Magazine,* November 2003, viewed at http://www.sierraclub.org/sierra/200311/media.asp; see also "Networks Gave Environment 4 Percent of 15,000 Prime News Hole Minutes in 2002," Metcalf Institute Environment Writer, June 2003, viewed at http://environmentwriter.org/resources/articles/0603_networks.htm.
8. See Federal Radio Act of 1927 at http://showcase.netins.net/web/akline/pdf/1927act.pdf.
9. See Fairness Doctrine, at http://www.museum.tv/archives/etv/F/htmlF/fairnessdoct/fairnessdoct.htm.
10. "Tobacco Advertising in the United States," Media Scope, March 9, 2000, viewed at http://www.mediascope.org/pubs/ibriefs/taus.htm.
11. *Friends of the Earth v. FCC,* 449 F.2d 1164 (D.C. Cir. 1971).
12. Bill Moyers, Keynote Address to the National Conference on Media Reform November 8, 2003, viewed at http://www.com mondreams.org/views03/1112-10.htm.
13. *Red Lion Broadcasting Co. v. FCC,* 395 U.S. 367 (1969), viewed at http://caselaw.lp.findlaw.com/scripts/getcase.pl?court=us&vol=395&invol=367.

14. *In re Complaint of Syracuse Peace Council against Television Station WTVH Syracuse, New York,* 99 FCC 2d 1389 (1984).
15. Mafruza Khan, "Media Diversity at Risk," *Corp Watch,* May 29, 2003.
16. Fairness Doctrine.
17. Eric Boehlert, "Is Clear Channel Playing a 'Shell Game'?" Salon.com, November 20, 2001, viewed at http://dir.salon.com/ent/clear_channel/2001/1½0/fcc_complaint/index.html.
18. William Safire, "Halt Media Madness," *Arizona Daily Star,* February 17, 2004.
19. "Super Polluters—The Top 25 Superfund Polluters and Their Toxic Waste Sites," Public Interest Research Group, viewed at http://www.pirg.org/reports/enviro/super25/index.htm.
20. Ibid.
21. "Goldman Environmental Prize Recipients," viewed at http://www.goldmanprize.org/recipients/recipients.html.
22. Complaint in *Steve Wilson and Jane Akre vs. New World Communications of Tampa, Inc.,* Florida 13th Judicial Circuit, General Civil Division.
23. "Florida Milk Supply Riddled With Artificial Hormone Linked to Cancer; Reporters Say They Were Ordered to Lie About It on Fox-TV," BGH Bulletin (2000), viewed at http://www.foxbghsuit.com/.
24. Michael Hansen, Ph.D., et al., "Potential Public Health Impacts of the Use of Recombinant Bovine Somatotropin in Dairy Production," September 1997, viewed at http://www.consumersunion.org/food/bgh-codex.htm.
25. "Dairy farmers applaud gov't rejection of growth hormone," CBC News, January 15, 1999, viewed at http://www.cbc.ca/stories/1999/01/15/sci-tech/bgh990115.
26. "New Zealand adopts strict tests to keep out genetically modified seeds," Associated Press Worldstream, August 1, 2002, viewed at http://ngin.tripod.com/010802b.htm.
27. Paul Elias/Associated Press, "Consumer Watch: Production cutback revives hormone debate; Tainted batches of Monsanto's engineered hormone used in dairy herds have rekindled safety concerns," *Philadelphia Inquirer,* February 15, 2004, p. E05.

28. "The Mystery in Your Milk," Ibid., viewed at http://www.foxbghsuit.com/exhibit%201r.htm.
29. Letter to Roger Ailes, CEO of Fox News, from attorney John J. Walsh, February 21, 1997, viewed at http://www.foxbghsuit.com/.
30. Ibid.
31. Complaint in *Steve Wilson and Jane Akre vs. New World Communications of Tampa, Inc.*
32. Steve Wilson, interview by Robert F. Kennedy, Jr., June 2004.
33. Complaint in *Steve Wilson and Jane Akre vs. New World Communications of Tampa, Inc.*
34. Trial transcript, *Steve Wilson and Jane Akre vs. New World Communications of Tampa, Inc.*
35. Steve Wilson, interview by Robert F. Kennedy, Jr., June 2004.
36. Opinion in *New World Communications of Tampa, Inc. vs. Jane Akre,* Florida Second District Court of Appeal, Case No. 2D01-529, February 14, 2003.
37. *New World Communications of Tampa, Inc. v. Akre,* 2003 WL 327505 (Fla.App. 2 Dist. 2003).
38. Opinion in *New World Communications of Tampa, Inc. vs. Jane Akre.*
39. *Amicus curiae* brief, *New World Communications of Tampa, Inc. vs. Jane Akre,* September 13, 2001, Florida Second District Court of Appeal, Case No. 2D01-529.
40. Order Granting Rehearing and Clarification and Denying Rehearing En Banc and Certification, *New World Communications of Tampa, Inc. vs. Jane Akre,* Florida Second District Court of Appeal, Case No. 2D01-529, February 25, 2003.
41. Steve Wilson, interview by Robert F. Kennedy, Jr., June 2004.
42. Jane Akre, interview by Robert F. Kennedy, Jr., June 2004.
43. Steve Wilson, interview by Robert F. Kennedy, Jr., June 2004.
44. Marvin Kitman, "Talk is Cheap, but Reruns are Cheaper Still, Chuck," *Newsday,* June 24, 1998, p. B43.
45. Arianna Huffington, interview by Robert F. Kennedy, Jr., April 2004.
46. Laurie David, interview by Robert F. Kennedy, Jr., April 2004.
47. Bill Moyers, "Bill Moyers on Big Media: The Lobbyists Overpower the People on FCC Ruling," *Public Affairs Television,* Oc-

tober 13, 2003, viewed at http://www.workingforchange.com/article.cfm?ItemID=15798.

48. Ibid.
49. Ibid.
50. John Dunbar and Aron Pilhofer, "Big Radio Rules in Small Markets" (part of the report "Well Connected"), Center for Public Integrity, October 1, 2003, viewed at http://www.publicintegrity.org/telecom/report.aspx?aid=63.
51. Moyers, "Bill Moyers on Big Media."
52. Ibid.
53. Ibid.
54. Bill Moyers, Keynote Address to the National Conference on Media Reform, November 8, 2003; see also Charles Layton, "News Blackout," *American Journalism Review,* January 2004.
55. Anne Marie Squeo and Joe Flint, "Court Bars Media-Ownership Rules," *Wall Street Journal*, June 25, 2004, p. A3.

11 | Reclaiming America

1. "Tax Administration: Comparison of the Reported Tax Liabilities of Foreign and U.S. Controlled Corporations, 1996-2000," April 2, 2004, Report #GAO-04-358, viewed at htttp://www.gao.gov/text/d04358.txt.
2. Address at the laying of the cornerstone of the Pilgrim Memorial Monument, Provincetown, Mass., August 20, 1907, viewed at http://www.time.com/time/time100/leaders/profile/troosevelt4.html.
3. "Military-Industrial Complex Speech," 1961, Public Papers of the Presidents, Dwight D. Eisenhower, 1960, p. 1035-104, viewed at http://coursesa.matrix.msu.edu/~hst306/documents/indust.html.
4. "Message proposing the 'Standard Oil' Monopoly Investigation," 1938, viewed at http://www.brainyquote.com/quotes/quotes/f/franklind135684.html.